Quality assurance workbook

for
radiographers &
radiological technologists

by
Peter J Lloyd
MIR, DCR, ARMIT, Grad Dip F Ed
Lecturer (retired),
School of Medical Radiation,
University of South Australia

Diagnostic Imaging and Laboratory Technology
Essential Health Technologies
Health Technology and Pharmaceuticals
WORLD HEALTH ORGANIZATION
Geneva

WHO Library Cataloguing-in-Publication Data

Lloyd, Peter J.
Quality assurance workbook for radiographers and radiological technologists / by Peter J. Lloyd.

1.Technology, Radiologic 2.Radiography 3.Quality assurance, Health care - methods 4.Quality control 5.Manuals 6.Teaching materials I.Title.

ISBN 92 4 154642 5 (NLM Classification: WN 180)

This publication is a reprint of material originally distributed as WHO/DIL/01.3

© **World Health Organization 2001, reprinted 2004**

All rights reserved. Publications of the World Health Organization can be obtained from Marketing and Dissemination, World Health Organization, 20 Avenue Appia, 1211 Geneva 27, Switzerland (tel: +41 22 791 2476; fax: +41 22 791 4857; email: bookorders@who.int). Requests for permission to reproduce or translate WHO publications – whether for sale or for noncommercial distribution – should be addressed to Publications, at the above address (fax: +41 22 791 4806; email: permissions@who.int).

The designations employed and the presentation of the material in this publication do not imply the expression of any opinion whatsoever on the part of the World Health Organization concerning the legal status of any country, territory, city or area or of its authorities, or concerning the delimitation of its frontiers or boundaries. Dotted lines on maps represent approximate border lines for which there may not yet be full agreement.

The mention of specific companies or of certain manufacturers' products does not imply that they are endorsed or recommended by the World Health Organization in preference to others of a similar nature that are not mentioned. Errors and omissions excepted, the names of proprietary products are distinguished by initial capital letters.

The World Health Organization does not warrant that the information contained in this publication is complete and correct and shall not be liable for any damages incurred as a result of its use.

The named authors alone are responsible for the views expressed in this publication.

Designed by minimum graphics

Printed in Malta by Interprint Limited

Contents

Introductory remarks	vii
Acknowledgements	viii
Introduction	**1**
Purpose of this workbook	1
Who this workbook is aimed at	2
What this workbook aims to achieve	2
Summary of this workbook	2
How to use this workbook	2
Roles and responsibilities	3
Questionnaire—student's own department	**5**
Pre test	**7**
Teaching techniques	**10**
Overview of teaching methods in common use	10
Assessment	10
Teacher performance	12
Suggested method of teaching with this workbook	12
Conclusions	12
Health and safety	**15**
Machinery	15
Electrical	15
Fire	15
Hazardous chemicals	16
Radiation	16
Working with the patient	17
Disaster	17
Module 1. Reject film analysis	**19**
Setting up a reject film analysis program	19
Method	20
Analysis	20
Action	20
Tasks to be carried out by the student	24
Module 2. Accessory equipment	**29**
Collimator	29
Cassettes and intensifying screens	31
Grid	35
Lead rubber aprons and gloves	37

QUALITY ASSURANCE WORKBOOK

Viewing box	38
Patient positioning aids	40
Patient measuring callipers	41
Tasks to be carried out by the student	44
Module 3. X-ray equipment	**56**
Choosing X-ray equipment	56
Acceptance of new X-ray equipment	57
Generator	57
X-ray tube, column, table and upright bucky	63
Tomography	65
Potter bucky	68
Portable and mobile X-ray units	70
Tasks to be carried out by the student	73
Module 4. Manual film processing	**78**
The darkroom	78
Film and chemical storage	81
Film processing	83
Tasks to be carried out by the student	95
Module 5. Automatic film processing	**107**
Choosing an automatic processor	107
Use of an automatic processor	108
Processor maintenance schedule	109
Sensitometry	112
Tasks to be carried out by the student	120
Module 6. Radiographic exposures	**124**
Exposure chart	124
The step system	126
Tasks to be carried out by the student	127
Appendix A. Making simple test tools	**133**
Water phantom	133
Aluminium step wedge	133
Film/screen contact test tool	134
Spinning top timer test tool	134
Measuring callipers	135
Tomography test tools	136
X-ray beam/grid alignment test tool	137
Appendix B. Graphs, check sheets and record sheets	**138**
Reject film analysis	138
Remedial action record	141
Collimator check records	142
Record of all cassettes	143
Cassettes and screens	144
Lead rubber aprons and gloves	146
Viewing box	147
Fault report	148
Equipment record	149
Equipment maintenance and repair log	152

X-ray unit	153
Constancy of radiation output at different mA settings	155
Darkroom inspection checklist	156
Specific gravity/temperature graph	157
Automatic film processor	158
Processor maintenance checklist	160
Developer activity chart	161
Characteristic curve chart	162
Quality control processing chart	163
Exposure chart	164

Post test   **169**

Glossary   **172**

References   **175**

Introductory remarks

This document, which is developed by the International Society of Radiographers and Radiological Technologists (ISRRT) under the umbrella of the WHO Global Steering Group for Education and Training in Diagnostic Imaging, is the first in a series targeting technical aspects, including quality control of diagnostic imaging services. The document is primarily aiming at assisting radiographers and radiological technologists working in small and mid-size hospitals where resources often are limited, to optimize and improve diagnostic imaging, and to ensure the best possible use of resources according to local needs.

The document is distributed free of charge and can be obtained by contacting the following address:

Team of Diagnostic Imaging and Laboratory Technology (DIL),
World Health Organization
20, Avenue Appia
CH-1211 GENEVA 27
Switzerland

Fax: +41 22 7914836
Tel: +41 22 7913648
e-mail: ingolfsdottirg@who.ch

Harald Ostensen, MD
Geneva, April 2001

Acknowledgements

Profound thanks are offered to everyone in Kenya who provided their wholehearted assistance, co-operation and enthusiasm in helping to produce this first workbook on quality assurance and pilot its implementation.

Thanks in particular to:

Kenya Ministry of Health
University of Nairobi
Nairobi Medical Training College
Kenya Association of Radiologists
Kenya Association of Radiographers
Kenyatta Hospital, Nairobi
and all other hospitals and government departments and individuals who participated in the initial research.

Thank you very much   Asante sana

Thanks also to all the people who assisted me in the production of this workbook:

Roger Windle, University of South Australia.
Rhonda Miller, University of South Australia.
Dean Hogben, Royal Adelaide Hospital.
Michael Fuller, Flinders Medical Centre.
David Lesley, South Australian Radiation Protection Branch.
Steven Johnson, South Australian Radiation Protection Branch.
Graham Truman Rtd, Adelaide Womens and Childrens Hospital.
Michael Canavan, Agfa-Gevaert Ltd.
Graham Blucher, Hanimex Pty Ltd.
Kenneth Gifkins, Kodak (Australasia) Pty Ltd.

Introduction

The World Health Organization (WHO) was founded in 1948 and is a specialised agency of the United Nations. It promotes technical co operation for health among nations, carries out programs to control and eradicate disease and strives to improve the quality of human life. WHO has four main functions:

- To give world wide guidance in the field of health.
- To set global standards for health.
- To co-operate with governments in strengthening national health programs.
- To develop and transfer appropriate health technology, information and standards.

The WHO definition of health

"Health is a state of complete physical, mental and social well-being and not merely the absence of disease or infirmity."

The International Society of Radiographers and Radiological Technologists (ISRRT) was formed in 1959 to act as a link between radiographers and radiological technologists throughout the world. The Society is dedicated to the improvement of standards of practice in radiation medicine technology. It is a non-political organisation and is not a trade union. It is a not-for-profit organisation.

The ISRRT is recognised as the international representative of radiation medicine technology throughout its official relationship as an international non-governmental organisation with WHO. It also has association with the United Nations and other organisations such as the International Society of Radiology, the European Society of Radiology, and the International Commission for Radiological Education.

WHO recognised the need to improve the standards of radiation medicine throughout all regions of the world. ISRRT, as part of this project, has entered into an agreement with WHO to produce a number of radiography technical workbooks, on various topics, for use in developing countries. This workbook on quality assurance is the first of these books.

Kenya was the country chosen to pilot the project.

Purpose of this workbook

It is preferred to call this a **workbook** rather than a manual or textbook, because the intent is to, not only give technical information, but to set practical exercises that students can work through, responding to specific questions. Above all, the students should feel that they have actually carried out the tasks themselves and will be more confident to teach others and ensure that these exercises continue to be carried out in their respective areas.

The topic of this workbook is quality assurance, so all material is designed to assist in the maintenance of the highest quality of work that can be achieved, under the prevailing conditions.

Quality assurance
The overall management program, put in place to ensure that a comprehensive range of quality control activities work effectively.

Quality control
The means by which, each area of interest is monitored and evaluated.

A Quality Assurance Program should be comprehensive, looking at all aspects of the work involved in producing high quality radiographs. This program should be cost effective and achieve its aims.

The ultimate responsibility for setting up, running, evaluating and taking remedial action lies with the head of department, although appropriate delegation may be necessary. It is important that someone accepts that responsibility and ensures that the program happens effectively.

This workbook will be used by radiographers who are being trained through a Centre of Excellence to:

- achieve a good working knowledge of quality control routines.

- return to their respective areas after completed training to teach other members of their staff to carry out the routines that they have learned.

In this way it is hoped that quality assurance will be practised routinely and effectively.

If so:

- Heads of departments will find that the standard of radiography will be maintained at the highest level.
- Work environments will be improved.
- Tasks will become easier.
- Repeat films will be kept to a minimum.
- Staff job satisfaction will increase.
- Patients will receive less radiation and less inconvenience.
- There will be fewer equipment failures.
- Costs will be kept down.
- A record and audit trail will exist as proof of high standards.

Achieve some of these and this workbook has been worth while!

Who the workbook is aimed at

This workbook is aimed directly at radiographic staff members of any X-ray department, who:

- are considered to have an adequate background and training in radiography,
- are concerned about the need to achieve the highest possible standards,
- have the ability and interest to learn and to teach others.

Indirectly this workbook is aimed at all radiographic and darkroom staffs through the selected member of staff chosen to undergo the initial training.

Heads of departments selecting the member of staff who is to undergo training should consider the personal attributes of their staff members and nominate the person who is most likely to achieve all of the aims.

What this workbook aims to achieve

- Increase awareness, interest and understanding of quality assurance issues.
- Enable radiographers to establish and continue to carry out an effective quality assurance program.
- Provide a comprehensive knowledge, advice and experience in quality control methods.
- Provide the knowledge and skills to carry out basic care and maintenance of imaging equipment.
- Raise standards.
- Reduce imaging costs.
- Improve job satisfaction.
- Improve health and safety issues.

Summary of this workbook

This workbook contains:

- Background information.
- A questionnaire seeking information about each students own department.
- A pre test of student's knowledge.
- Advice on teaching methods.
- Health and safety issues.
- Six modules giving technical information regarding effective routines and quality control techniques.
- Each module contains relevant tasks the student must perform.
- How to make simple test tools.
- Copies of quality control documentation.
- A post test of student's knowledge.
- Glossary of terms.
- Reference list.

How to use this workbook

Once a radiographer has been selected and training dates set, this workbook should be made available to the student at least *two weeks* in advance of the commencement of the training period, so that adequate pre reading and discussion can be carried out.

The student must be encouraged to discuss the content of the workbook with other members of staff during this pre reading period.

The section headed STUDENT'S OWN DEPARTMENT, must be *completed by the student before commencement of the course*. This takes the form of a questionnaire which, when completed should give the tutor a background knowledge of the student and their work environment. This background information will allow the tutor to apply the correct emphasis when providing and supervising the training.

The student must complete a PRE TEST prior to starting the course. This is an assessment of the student's relevant knowledge *before* the course. This will be compared to the results of a similar POST TEST completed by the student *after* completion of the course. These tests are for student information and course evaluation only and are *not* used in student assessment.

The section on TEACHING TECHNIQUES first gives a broad overview of teaching methods. This is followed by the *recommended* approach to teaching with this workbook. *Both tutor and student should read this section.*

The section on HEALTH AND SAFETY draws attention to all the health and safety issues appropriate to an X-ray department and how to make the work environment a safe and healthy one.

The workbook is divided into modules.

- The student should work through one module at a time, studying the technical information and testing methods.
- At the end of each module, tasks have been set. The student must carry out each task and answer the questions asked.
  — The tutor will assess the completed **tasks and answers**, adding any appropriate comments.
  — A **Satisfactory/Unsatisfactory** grade will be given.
  — All **Unsatisfactory** exercises must be repeated before beginning the next module.

All necessary equipment will be provided by the Centre of Excellence. The tutor will ensure that the student understands the technical information given in the workbook and will supervise and advise during the practical exercises.

The APPENDICES contain information on making **simple test tools, report forms, record sheets, test result sheets** and **exposure charts**, for use in the student's own department.

The GLOSSARY contains a list of **terms**, found in the text, with meanings.

The REFERENCES provide a source of further reading.

The POST TEST must be answered on completion of the course and the workbook handed to the tutor for final assessment and handing on to the Head, Centre of Excellence, for final approval.

The workbook will then be handed back to the student who will use it, on returning to their own department, to train other members of staff and to ensure that a quality assurance program is established and carried out on a regular basis.

Roles and responsibilities

The Head, Centre of Excellence

The Head, Centre of Excellence will:

- Hold the overall responsibility for the organisation and presentation of the training program related to this workbook.
- Be responsible for the selection of suitable tutors and ensure that they are fully trained and aware of their obligations.
- Ensure that all necessary facilities and equipment are available.
- Arrange for workbooks to be in the hands of students at least *two weeks* before they start their course, for pre reading and discussion with colleagues.
- Ensure that all training related to this workbook is carried out satisfactorily.
- Receive, review and sign all completed workbooks and return them to the tutor for onward transmission to the student.
- Take any necessary action arising from completed workbooks, student performance or student behaviour.
- Give feedback to the tutor.
- Arrange for follow up checks to be carried out, after a suitable period of time, on each student and what they have achieved in their own department since completing the course.
- Evaluate follow up reports and take any necessary action.

The tutor, Centre of Excellence

Each student will be allocated a tutor from the staff of the Centre of Excellence. The tutor will:

- Be responsible for students to whom they have been allocated.
- Be readily available to students during their training.
- Supervise and teach students during their training.
- Familiarise his/her self with the workbook, in particular the section on teaching techniques.
- Formulate a strategy for teaching this course and implement it.
- Read the questionnaire STUDENT'S OWN DEPARTMENT, completed by the student, and devise an appropriate training program.
- Ensure that all necessary equipment is available.

- Ensure that the training program is carried out.
- Evaluate all tasks carried out by the student and write appropriate comments and a grading at the end of each task sheet.
- Ensure that tasks graded "Unsatisfactory" are repeated before the student progresses to the next module.
- Mark pre and post course tests, completed by the student, and make the student aware of the results.
- Submit the student's completed workbook to the Head, Centre of Excellence for final approval.
- Return the workbook to the student.
- Give adequate feedback to the student.
- Ensure that the student is fully aware of their responsibility to teach the course topics to fellow staff members on returning to their own department.
- Discuss with and advice students on how to carry out their own training programs.
- Advise the student that a follow up check will be made to assess the benefits of the course.

The student

The student will:

- Complete all pre reading, discuss the material with colleagues and fill in the questionnaire, STUDENT'S OWN DEPARTMENT, before starting the course.
- Carry out the PRE TEST immediately before starting the course.
- Attend all scheduled teaching, practical and administrative sessions.
- Complete all modules, by first reading the technical information, carrying out the allotted tasks under the supervision of their tutor, and then answering the questions.
- Submit the workbook to the tutor, for evaluation, upon completion of each task.
- Repeat any task assessed as "Unsatisfactory" before progressing to the next module.
- Carry out the POST TEST immediately upon completion of the course.
- At the end of the training period, and when all tasks have been satisfactorily completed, hand the workbook to their tutor for final evaluation and submission to the Head, Centre of Excellence for formal approval. The workbook will be returned.
- On return to their own department, use the workbook and newly gained knowledge and expertise to establish a quality assurance program and train colleagues, under the direction of their Chief Radiographer.

Questionnaire
Student's own department

In order for this course to meet your needs, your tutor must know something about the department in which you work. Please answer the following questions in the spaces provided, before you commence the course.

1. How many X-ray examination rooms are there? _____

2. Tell us what X-ray equipment you have. e.g. general purpose table with bucky etc.

   ROOM 1 _____

   _____

   _____

   ROOM 2 _____

   _____

   _____

   ROOM 3 _____

   _____

   _____

3. Tell us what accessory equipment you have. e.g. Cassettes, positioning pads etc.

   _____

   _____

4. How many darkrooms are there? _____

5. State the type of film processor in each darkroom. e.g. manual/auto, type, make, model, processing cycle.

   DARKROOM 1 _____

   _____

   _____

   DARKROOM 2 _____

   _____

   _____

QUALITY ASSURANCE WORKBOOK

DARKROOM 3

6. How many staff act as radiographers? Qualified radiographers _____ Others _____

7. Is there an "out of hours service"?   YES/NO

8. How many darkroom technicians are there? _____

9. Do you already run any form of Quality Assurance Program?   YES/NO

10. If YES state here what you do

11. List any quality control test tools you have

12. If you have any quality assurance issues you particularly want covered, state them here

13. Give an indication of how busy your department is. Number of examinations per year/week/day _____

14. Summarise the types of examinations carried out. e.g. extremities, chest, spine.

Thank you

Pre test

This test must be completed by the student *before* starting the course. The intention is to test your knowledge on the topics covered by this workbook *before* the course.

You will be tested again when you have completed the course so that you have an idea of how much you have learned.

The results of these tests are for information only and will not affect your course result.

Instructions

This is a multiple choice test. In each question you are given three possible answers.

Read each question carefully.

Indicate the answer that you feel is the most accurate by placing an "X" in front of the letter preceding it.

Example:
A personal radiation monitor (TLD) should be worn
    a) Outside a lead rubber apron.
X b) Under a lead rubber apron.
    c) There is no need to use one when wearing a lead rubber apron.
Answer: b)

All questions must be answered

1. What is meant by the term "quality assurance"?
    a) The equipment is covered by an insurance policy.
    b) Everyone must produce perfect films.
    c) A system which attempts to maintain a high quality of work all round.

2. What is meant by the term "quality control"?
    a) A practical exercise which carries out quality checks.
    b) A staff member who supervises quality.
    c) An X-ray machine system that gives the correct exposure.

3. Reject film analysis means?
    a) To ask the radiographers how many films were repeated that day.
    b) A detailed study of film wasted over a period of time.
    c) Counting all the films accidentally fogged in the darkroom.

4. A grid ratio is ?
    a) Ratio of the width of a grid to its length.
    b) Ratio of the height of the lead strips to their length.
    c) Ratio of the height of the lead strips to the distance between them.

5. A stationary grid is?
    a) A grid that is fitted in a bucky.
    b) A grid that can be carried around.
    c) A series of shelves for filing papers.

6. To test for poor film screen contact:
    a) X-ray a lot of paper clips on the face of the cassette.
    b) Open the cassette and look.
    c) Place a sheet of fine wire mesh inside the cassette with the film and make an exposure.

7. To check if the light field and the X-ray field of a collimator are correctly aligned:
    a) Look into the collimator mirror.
    b) Open and close the collimator shutters rapidly.
    c) Place metal markers on the face of a loaded cassette to indicate the light field and make an exposure.

8. The coincidence of the X-ray and light fields of a collimator are said to be acceptable when:
    a) The X-ray field is 15 mm inside the light field at 100 cm FFD.
    b) The X-ray field is 3 mm outside the light field at 100 cm FFD.
    c) The X-ray field is 8 mm inside the light field at 100 cm FFD.

9. The wattage of a light bulb in a darkroom safelight facing down should be:
    a) 15 watts.
    b) 50 watts.
    c) 100 watts.

10. Static electricity:
    a) Produces an overall grey fog on processed film.
    b) Produces black lightning like marks on processed film.
    c) Reduces the effect of the intensifying screens.

11. A densitometer:
    a) Accurately assesses film density.
    b) Determines the efficiency of lead rubber.
    c) Determines the light output of intensifying screens.

12. A Sensitometer:
    a) Is a motion detector.
    b) Is a radiation detector.
    c) Is a device for making test strips used in film processor monitoring.

13. Fixing time in manual processing should be:
    a) Two minutes.
    b) Twice the clearing time.
    c) Twenty minutes.

14. X-ray film should be:
    a) Stored at a temperature of 10° to 20°C.
    b) Stored lying flat.
    c) Handled only in total darkness.

15. Stock rotation, related to film storage, means:
    a) Turning the film boxes around.
    b) First in first out.
    c) First in last out.

16. A characteristic curve:
    a) Determines the shape of an object.
    b) Represents the characteristics of the developer.
    c) Is a graphical representation of the relationship between the exposure received by the film and the density produced, following processing.

17. Safelights facing down should be installed:
    a) No less than 130 cm above the workbench.
    b) No less than 100 cm above the workbench.
    c) At least 150 cm above the workbench.

18. Automatic processing developer temperatures should be:
    a) 20°C.
    b) 25°C.
    c) 35°C.

19. In manual processing the film should be:
    a) Agitated every 30 seconds in the developer.
    b) Placed in the developer and not touched until the timer sounds.
    c) Agitated by moving the film sideways.

20. In manual processing, when transferring the film from the rinse to the fixer, the film should be drained into:
    a) The rinse.
    b) The fixer.
    c) Doesn't matter.

21. In addition to a visual safelight check you should:
    a) Hold an unexposed film against the safelight for one minute.
    b) Expose sections of a film to safelight for progressively longer times.
    c) Stand in the centre of the darkroom holding a film for one minute.

22. If the lid has accidentally been left off a box of unexposed films in white light:
    a) Throw all the films away.
    b) Put the lid back on and do nothing until someone tells you their films are fogged.
    c) Process three films and inspect.

23. The First Aid treatment for a processing chemical splash in the eye is:
    a) Blink continuously for 30 seconds.
    b) Wash thoroughly.
    c) Wipe the eye with a tissue.

24. The test to determine optimum development time is:
    a) Develop test strips for differing times and compare.
    b) Develop an unexposed film and inspect.
    c) Develop test strips for the same length of time.

25. When shutting down an automatic processor:
    a) Clean crossovers and leave lid partially lifted.
    b) Switch off and leave it.
    c) Turn water off only.

26. Developer temperature should be checked:
    a) Only when film densities look different.
    b) Once a week.
    c) Daily.

27. Replenishment rates in automatic processors are checked by:
    a) Catching the amount pumped in a graduated flask.
    b) Asking the manufacturer.
    c) Measuring the drop in level of the replenishment bottle/tank.

28. To check that a generator always gives the same output when using the same exposure factors:
    a) Watch the mA meter during the exposure.
    b) Expose three separate areas on the same film, using the same exposure factors.
    c) Expose three different cassettes using the same exposure factors.

29. A spinning top test is used to:
    a) Check the accuracy of the timer.
    b) Check the accuracy of the mA.
    c) Check the accuracy of the kV.

30. To test for constancy of radiation output of an X-ray unit:
    a) Make several exposures using a different kV each time.
    b) Make three exposures keeping the kV the same, but varying the mA.
    c) Make three exposures keeping the kV and mAs the same but varying the mA and time.

Teaching techniques

How you teach a subject is very important. The most highly skilled and knowledgeable person can fail to pass on the necessary information to students by not using the correct teaching skills or methods.

Following is a broad coverage of teaching methods, some of which will be appropriate to your situation and others not.

You should consider how you may perform and what method you will use. Choose a method suited to your subject and what you aim to achieve.

Planning is all-important. Research your topic well.

Overview of teaching methods in common use

Methods of presentation

Lecture
- Stand or sit in front of a class and verbally give the relevant information.
- Suitable for large and small classes.
- Rather inflexible.
- Can be boring.
- Audio visual aids can be used.
- Printed notes can be given out in support of the spoken word.

Tutorial
- More informal than the lecture.
- Suitable for smaller classes.
- Students are encouraged to present material and enter into group discussion.
- The teacher acts more as a facilitator.
- Participants can sit around a table or in a circle.
- Feedback is important.

Practical
- Students carry out a practical exercise under teacher supervision.
- Verbal or written instructions may be made available to the student.
- The practical should be relevant to recently acquired information.

Demonstration
- Given by the teacher to illustrate a particular point.
- May be carried out as a supplement to a lecture or tutorial or an introduction to a practical.

Role Play
- Students act out specific sets of circumstances.

Reading
- Students are given topics or specific references to read up on.
- Often used as a preliminary before a tutorial or instead of a lecture.

Self directed learning
- The student is given the topic and the expected outcome.
- The student does their own research and problem solving.
- Exchange of information and group problem solving is encouraged.
- Usually followed up by a tutorial and a written confirmation of the student's knowledge.

Context based learning
- Problem solving, in groups.

Presentation
- The student researches the topic and gives a talk to other students.
- The teacher acts as facilitator and assessor.

Assignment
- The student researches a set topic and hands up a written presentation to the teacher.

Workbook
- A specialised book that poses questions and/or sets practical tasks.
- Answers are generally recorded in the book.

TEACHING TECHNIQUES

Book review
- To encourage students to read certain books.
- The student is asked to present a written summary and comment on the book.

Posters
- Can be done in work groups.
- May be used as teaching material at a later stage.
- Could be displayed in department to convey information.

Presentation technique

In any form of presentation by a teacher to a class, the following points should be considered:

- Stand or sit where students can see you.
- Address them clearly.
- Use language they can understand.
- Present your facts logically.
- Speak directly to the class.
- Cast your eyes around the class as you speak.
- Use visual aids where necessary.
- May use flip charts.
- In a tutorial or practical situation the teacher acts largely as a facilitator.

Teaching aids
- 35 mm slides.
- Overhead projection (OHP).
- Video
- White Board / chalk Board
- Butchers paper.
- Models,
- Charts.
- Radiographs.
- Pieces of relevant equipment.
- Printed notes.

Preparation
- Select the topic.
- Research the topic.
- Assess the educational and technical level of the students.
- Decide on the breadth and depth of the material to be covered.
- Select the method of teaching to be used.
- Select the teaching environment.
- Prepare relevant teaching notes.
- Prepare relevant teaching aids.
- The length of the session must match the time slot available.
- The amount of material to be presented must match the length of the time slot.
- In the case of a practical, ensure before hand that it will work.

Running a practical
- Identify each piece of equipment.
- Explain the procedure.
- Outline the aim.
- Demonstrate if necessary.
- Identify likely problems.
- Observe student carrying out the task.
- Comment as necessary.
- Be available to assist or answer questions.

Feedback to the student

Following any learning activity performed by the student, the teacher should give feedback to the student regarding their performance.

- Mark / grade achieved.
- Method.
- Content.
- Technique.
- Performance.
- Presentation.
- Or whatever is appropriate.

Assessment

- Written examination.
  - Multiple choice.
  - Short answer.
  - Essay.
  - Numerical answer.
  - Problem solving.
  - Open book.
  - Fill in a missing word.
  - Tick a box.
- Practical task (written question and answer).
- Practical task (teacher observed—must have a pre-determined mark sheet).
- Assignment.
- Book review.
- Oral examination.

Grading student work

- Written.
- Verbal.
- Practical.

Forms of grading
- Pass / Fail / Referred.
- A B C D E F
  - Where A is the highest and F is the lowest.

- — A to E are graded passes.
- — E to F are graded fails.
- Distinction / Credit / Pass / Fail.
- Marks out of 100.
- Marks out of 10.
- Combined project(s)/exam(s) e.g. 50%/50% or 25%/25%/50%
- Satisfactory/Unsatisfactory.
- No grading, the student simply attends classes and completes all work set.
- Outcome based
- Can the student perform a task satisfactorily.
- The student must be made aware of the grading system.
- The student must be made aware of the results of any assessment.
- The teacher must allocate marks to the various parts of the material before marking Takes place.

Teacher performance

- Start and finish on time.
- Plan content to fit the time frame.
- Encourage questions.
- Speak clearly.
- Maintain a friendly discipline.
- Be well prepared:
  - — Knowledge.
  - — Notes.
  - — Teaching aids.
- Mark all work fairly and accurately.
- Return all marked work as soon as possible.
- Give feedback.
- Regularly look around your audience.
- Humour is a useful tool if used properly.
- Don't be sexist, superior, aggressive or condescending.

How good are you at teaching?

Some ways of understanding how you perform as a teacher:

- Evaluation questionnaire filled in by the student at the end of a class or series of classes.
- Ask someone to watch you teach and give you feedback.
- Watch student reaction during a class.
- Videotape a lecture, view it and then evaluate yourself.

Sample evaluation questionnaire of teacher performance

- Were you able to hear? YES / NO
- Did the teacher start and finish on time? YES / NO
- Was the material presented in a logical way? YES / NO
- Was the material covered adequately? YES / NO
- Was the presentation carried out satisfactorily? YES / NO
- Was the attitude of the teacher satisfactory? YES / NO
- Were the aim and objectives of the course achieved? YES / NO
- Further comment:

_____

_____

- Instead of YES / NO the student could be asked to choose from the following:
  - — Excellent.
  - — Good.
  - — Acceptable
  - — Needs improvement

Conclusion

- This section has set out to give you an overall insight into teaching.
- Much of the material given will not directly apply to your situation.
- You must choose what you feel is relevant to you and carry out your teaching to suit your own needs and those of the student.

Suggested method of teaching with this workbook

This book sets out to give the relevant information on the topics listed in the contents, regarding quality assurance, then sets simple tasks that students are expected to carry out, responding to any questions asked. The teaching method used should therefore be practically oriented.

The tutor

A tutor using this book must:

- Become thoroughly familiar with the book.
- Understand its contents and know how to perform all tasks set.

- Know the answers to all questions asked.
- Understand and be able to run practical exercises and carry out any other form of teaching considered appropriate.
- Be able to select and use the most appropriate method of teaching.
- Make sure that all equipment needed is available.
- Make sure the student understands what is required of them.
- Be available to advice while the student is carrying out the exercises.
- Assess student's work fairly and accurately.
- Give useful feedback.
- Be sympathetic to student needs.

Method

It is suggested that a tutor use the following basic teaching format:

- Understand the **student's needs**.
- Devise an appropriate **teaching program** (lecture, tutorial, practical etc.).
- Ensure that you have all the **equipment and teaching aids** that you require.
- You should carry out a practical exercises yourself first to ensure it works.
- Ensure that you have the correct **answers/results**.
- Issue the workbook to the student, at least two weeks **before** they start the course, for pre reading discussion with colleagues and completion of the information about their department.
- Read the completed questionnaire, **Student's own department**, and discuss with the student in order to determine their needs.
- Before starting the course the student *must* complete the **Pre test**.
- Outline the **teaching format** to the student.
- Identify the **topics** to be covered.
- Give a copy of the **teaching program** to the student.
- Cover one topic at a time.
- With each topic, first give formal instructions covering all the **relevant information**.
- Answer any **questions**.
- Allow the students to carry out **the exercises**.
- The tutor should be available in an **advisory capacity**.
- Give the student **time to complete the exercise** and any necessary written work.
- **Assess** the practical component.
- **Assess** the answers on the task sheet and make written comment.
- **Grade** the answers and practical performance **Satisfactory/Unsatisfactory**.
- Give **feedback** to the student.
- If the student's performance is **unsatisfactory** the task must be repeated.
- Continue with the next topic when a **satisfactory** grade has been achieved.

Notes

Health and safety

Health and safety issues in any work environment are very important. **It is the responsibility of all Heads of Department to ensure that injury or sickness, due to working conditions, is kept to a minimum. Injury or sickness may increase absenteeism of staff members and reduce efficiency. Staff must not put patients, colleagues or self at risk.**

X-ray departments should be prepared for emergencies such as fire, major disaster or any life threatening situation. Radiography involves working with:

- Machinery.
- Electricity.
- Hazardous chemicals.
- Radiation.
- Patients.

Fire and major disaster plans should be in place. Cardiac arrest training should be given.

The first thing to consider is making the work environment as safe as possible, by minimising the risk of problems arising. To achieve this, ensure that:

- Regular maintenance inspections are carried out.
- Safety procedures are followed.
- Adequate staff instruction is given.
- Safety equipment is readily available.

Machinery

Regularly inspect all machinery. Do not attempt to repair anything you do not understand. Call an X-ray engineer if you are unable to fix the problem.

Take care with all moving parts, to minimise the risk of:

- Trapping fingers.
- Loose parts falling off onto staff or patient.
- Equipment moving unexpectedly and striking staff or patient.
- Staff or patient striking head on overhead equipment.

Electricity

Consult a qualified electrician or X-ray engineer. Regularly inspect all electrical equipment, cables and connections. **Do not attempt to repair anything you do not understand.**

When carrying out any simple maintenance, repair, or cleaning of electrical equipment:

- Switch off and disconnect before starting.
- Do not tamper with anything you do not fully understand.
- Unless you are qualified, restrict your actions to replacing light bulbs, simple electrical parts, tightening connections, replacing fuses and inspecting cables.
- Ensure that all parts are correctly and safely installed or adjusted.
- Ensure that all protective panels are replaced.
- Never use excessive amounts of water when cleaning electrical equipment.
- Report all faults to your immediate senior or through the recognised channel.
- Ensure that other members of staff are aware of any problem.

Fire

Adequate fire fighting equipment, instructions, and evacuation procedures must be in place at all times.

Equipment
- Fire extinguishers.
  - Suitable for electrical fires, near electrical equipment and switchboards.
  - General purpose in all other areas.
- Smoke detectors in all rooms.
- Hoses in central areas.
- Fire alarms easily accessible.
- Regular maintenance of alarms and equipment.
- Illuminated EXIT signs in all public area.
- Emergency exit doors not locked or blocked.

Fire fighting instructions
- Readily available.
- Staff training.
- Annual refresher courses.

Evacuation procedure
- Instructions readily available.
- Clearly defined evacuation routes.
- Recognised assembly points.
- Responsibilities clearly defined.
- Staff training.

Hazardous chemicals (laws and regulations to be followed)

Developer and fixer are hazardous chemicals and should be handled with care. Display manufacturer's instructions for mixing, care and first aid treatment, in a prominent place in the area in which the chemicals are to used.

The risks involved are:
- Inhaling fumes or powders.
- Swallowing.
- Contact with the skin or eyes.

When mixing solutions:
- Work in a well-ventilated room.
- Avoid skin or eye contact with chemicals.
- Wear a mask, goggles, rubber gloves and a plastic apron.
- Avoid splashes.
- Wash all equipment used after mixing.
- Clean up any spills or splashes.

When processing films:
- Avoid skin or eye contact with chemicals.
- Ensure that the darkroom is adequately ventilated.
- Minimise splashes.
- Clean up any splashes as soon as possible.
- Replace any tank lids when finished.

Disposal of empty chemical bottles
- Should not be used as drinking water containers.
- Puncture and place in a sealed plastic bag before disposal.

Disposal of exhausted chemistry. Things NOT to do.
- Do not flush into common drains or simply throw away. The chemicals may get into the local water supply or contaminate crops.
- Do not flush into a septic tank system. The chemicals will kill the "good" bacteria and stop the breakdown of solid matter.

Disposal of exhausted chemistry. Helpful suggestions.
- Ideally use a silver recovery unit and dispose of the chemistry through a recognised hazardous chemicals agency.
- Select a suitable site where the chemicals can be buried and are not likely to get into the local water supply or in any way affect humans, animals or crops.
- Further refinements of the "bury method" is to use a sand trap first, then bury the residual sand or use an evaporative trench lined with sand and bury the sand when the water has evaporated.
- Local soil, terrain and weather conditions should be considered.

First aid treatment
- Follow manufacture's recommendations.
- Skin contact.
  — Wash thoroughly in water immediately.
- Eye contact.
  — Wash eye thoroughly, immediately.
  — Darkrooms should be equipped with emergency eye wash kits.
- Inhaled
  — Move out into fresh air immediately.
  — Seek medical advice.
- Swallowed
  — Wash mouth and lips in clean water.
  — Seek medical advice immediately.

Radiation

Follow national laws and regulations.
- Use an ongoing personal monitoring system.
- Do not produce x-radiation unnecessarily.
- Requests for X-ray examinations should be justified.
- Avoid the use of X-ray examinations on pregnant women wherever possible, especially in the first trimester.
- Keep clear of the primary radiation beam.
- Keep clear of any scattered radiation.
- Collimate the beam as much as practicable.
- Minimise repeat films.
- Use lead rubber shielding whenever possible, especially of radiosensitive organs.
- Make sure that all items of lead rubber are in good condition and effective.
- Make sure that shielding to the control panel is effective.
- Make sure that X-ray room walls effectively protect people in adjacent areas.
- Close door to X-ray room when exposing.
- Standard radiation warning symbols must be placed on the doors of all X-ray rooms.
- Illuminated signs should be placed at the entrance to all X-ray rooms where prolonged X-ray

exposures are made, warning when X-rays are being used. e.g. screening rooms.
- Make sure that all unnecessary personnel are clear of the radiation area when exposing.
- Make sure that X-ray equipment is working properly and is safe, by carrying out regular quality control checks.
- X-ray equipment should be switched off when not in use and any safety lock keys removed.
- Use correct filtration of the X-ray beam.
- Special care must be taken when using mobile/portable X-ray units, in ward or operating theatre situations.
- Apply the ALARA or ALARP Principle when exposing anyone to radiation.

ALARA principle
When exposing anyone to ionizing radiation the dose should be kept
As Low As Reasonably Achievable

An extension of this principle is the

ALARP principle
When exposing anyone to ionizing radiation the dose should be kept
As Low As Reasonably Practicable

The inference here is that there may be other important factors which may limit the dose reduction.

Working with the patient

Moving and handling:
- Use recognised moving/handling techniques to reduce the risk of back injury.
- Use appropriate moving/handling aids when necessary and when available.
- Encourage patients to move themselves where possible.

Cross infection:
- Be aware of any indications that the patient may be infectious.
- Follow infection control procedures.
- Have gowns, gloves and masks readily available and ensure that they are worn when necessary.
- Disinfect all equipment used, immediately after use, not forgetting cassettes.
- Where possible wrap cassette in towel or cloth before using, with an infectious patient.
- Wash hands before and after each contact.

Disaster

A disaster is any major catastrophic event, earthquake, major accident involving many people, civil disturbance or war. A hospital must have an established disaster plan that will allow it to cope with this type of event. An X-ray department must have its own emergency plan that will fit in with the overall hospital plan.

Designing a plan:
- Become familiar with the hospital plan.
- Design an X-ray department plan that is compatible with that of the hospital.
- Determine:
    — Individual responsibilities.
    — Staff call-out procedure.
    — Organisation.
    — Amount of stock to hold.

Notes

MODULE 1
Reject film analysis

Aim

To provide the knowledge necessary to set up a system which will give a detailed analysis of the rejected films and the reason for the rejections, in order to put in place remedial action.

Objectives

To provide all relevant information and give the opportunity to gain practical experience in:
- Designing a reject film analysis program.
- Setting up and running a reject film analysis program.
- Analysing the results of that program.
- Setting up remedial action based on those results.
- Instigating a follow up program to assess the effectiveness of the remedial action taken.

In every department a number of films are discarded for one reason or another. Incorrect exposure, poor positioning and processing are some of the common causes.

Knowing exactly what the major reasons are is a big step towards correcting the faults and therefore reducing the number of unacceptable radiographs. It is not satisfactory to make judgements on impressions alone. A reject analysis can be an important part of your ongoing quality assurance program.

From a radiographers point of view a reject analysis may be seen as a threat. It is important to fully explain to all concerned:

- Why the analysis is being carried out.
- How it will be carried out.
- When it will be carried out.
- The benefit.

Sensitometry and radiation consistency tests should be run in parallel with a reject film analysis as processing or radiation output irregularities may be causing some of the problems.

Benefits
- Able to identify the main errors and put measures in place to reduce them.
- Save money by reducing wastage.
- Reduce radiation dose to the patient by minimising the number of repeat films.
- Save time and effort by reducing the number of repeat films.
- Provide ongoing data for comparison.
- Provide possible source of statistics to support claims for more funding to replace, modify or repair faulty equipment.

Potential problems
- Staff members do not co operate fully.
- Radiologist and clinicians tend to retain substandard films on the grounds that they provide some information.
- Reject film records not always kept up to date.

Frequency
- Monthly.

Setting up a reject film analysis program

- Design the program.
- Determine the period over which the program will run.
- Nominate a starting and finishing date and time.
- Inform staff (*the full co operation of the staff is important*):
  — Why the program is necessary.
  — How it will operate.
  — When it will take place.
  — Who is responsible for the program.
- Decide what data you need.
- Design data recording sheets.
- Place reject film boxes in appropriate areas.

It is usual to collect reject films from individual X-ray rooms. *It may be necessary to relate the collections to the individual radiographer as well.*

- Immediately before program starts:
  — Count all films in store by sizes.
  — Add to this count all unexposed films in cassettes, film hoppers and partially empty boxes.
  — Record this information.
  — Dispose of all current reject films.

Method

- Collect all reject films daily.
- Record numbers of reject films on data sheets daily (see Fig 1–1, Fig 1–2, Fig 1–3 and *Appendix B*, pages 138–140).
- Immediately the program has finished re-count the films in the store, cassettes and half empty boxes.
- Original film count, minus end film count, equals number of films used.
- Analyse the data.

Analysis

The following information can be obtained from an analysis of the data:

- Overall number of reject films.
- Number of reject films by size.
- Number of reject films by faults.
- Number of reject films by rooms (or radiographers).
- Overall cost of rejected films.
- Identification of common faults.
- Reject films as a percentage of films used.

Study the data. Compare current data with that of previous programs.

A reject rate of 10% or more should be considered unacceptable. A reject rate of 5% to 10% justifies continued monitoring.

Action

- Rank the faults in order of frequency.
- Identify the most common faults.
- Set up remedial programs to correct errors (see *Appendix B*, page 141).
- Inform staff of findings and planned remedial action.
- Start remedial action program.
- Nominate dates for another analysis if felt necessary.
- File data for future reference.

Reject film analysis
Daily totals by location

Date:

Cause	Location				
	Room 1	Room 2	Room 3	Room 4	Total
Projection					
Movement					
Under exposed					
Over exposed					
Static					
Fog (darkroom)					
Fog (cassette)					
Equipment					
Processing					
Other					
Total					

Total films rejected today: _____

Fig 1–1. Reject film analysis, daily record sheet by location

Reject film analysis
Daily totals by film size

Date:

	Film size (cm)	No. of films	Film used (sq m)
1	35 x 43		
2	35 x 35		
3	30 x 40		
4	24 x 30		
5	18 x 24		
6	18 x 43		
7	DENT / OCC		
8	Other		
	Total		

Total number of films: _____

Total square meters of film: _____

Total cost of rejected film today: _____

Fig 1–2. Reject film analysis, daily record sheet by film sizes

Reject film analysis
Daily totals

Period: _____ From: _____ To: _____

Cause	Dates																Total
Projection																	
Movement																	
Under Exp.																	
Over Exp.																	
Static																	
Fog (darkroom)																	
Fog (cassette)																	
Equipment																	
Processing																	
Other																	
Totals																	

Total films rejected: _____

Total films used: _____

Rejected films as a percentage of films used: _____

Cost of films rejected this period: _____

Fig 1–3. Reject film analysis daily totals sheet

Notes

TASK 1
Reject film analysis

You have become aware that the number of rejected films is increasing, but you have no idea what the main faults are and what is causing them. You decide to investigate the situation. Take the first step and practise analysing films for faults.

  a) Select any three rejected films.
  b) Study these films.
  c) List all faults seen, in order of importance, that you believe led to the rejection of the film.

FILM 1	FILM 2	FILM 3

1. How do you think these faults were caused?

   Film 1 _____

   _____

   _____

   Film 2 _____

   _____

   _____

QUALITY ASSURANCE WORKBOOK

Film 3 _____
_____
_____

2. What would you do to eliminate these errors?

Film 1 _____
_____
_____

Film 2 _____
_____
_____

Film 3 _____
_____
_____

Tutor's comments:

Satisfactory/Unsatisfactory

Signed _____ Date _____
Tutor

TASK 2
Reject film analysis

Your next step is to carry out a small pilot reject film analysis.

    a) Carry out a *one room* reject film analysis over a period of *5 to 6 days* (see notes).
    b) Analyse your results.
    c) Answer the following questions in the spaces provided.

1. Total number of reject films: _____

2. Reject films as a percentage of total films used: _____

3. Total area in square metres, of rejected film: _____

4. Cost of rejected film? _____

5. List the three main causes of film rejection:

   _____

   _____

6. Explain what measures you would put in place to eliminate these causes:

   _____

   _____

   _____

   _____

   _____

   _____

   _____

QUALITY ASSURANCE WORKBOOK

Tutor's comments:

Satisfactory/Unsatisfactory

Signed　　　　　　　　　　　　　　　　　　　　　　　Date

Tutor

MODULE 2
Accessory equipment

Aim
To provide the knowledge and practical skills necessary to establish and carry out a quality control program for accessory equipment.

Objectives
On completion of this module you will be able to understand and carry out quality control procedures, evaluation and recommend action for:

Collimator
- Visual inspection.
- Accuracy of scales.
- Change a light bulb.
- Light beam/X-ray beam alignment and centring.
- Lead shutter efficiency.

Cassette/intensifying screens
- Visual inspection.
- Light tightness of cassette.
- Film/Screen contact.
- Intensifying screen light colour emission.
- Care and cleaning.

Grid
- Grid types, ratio and line.
- Care and cleaning.
- Visual inspection.
- Grid line damage.

Lead rubber apron, gloves and sheets
- Visual inspection.
- Cracking.
- X-ray testing.
- Care and cleaning.
- Storage.

Viewing box
- Electrical inspection.
- Efficiency/consistency.
- Location.
- Cleaning.

Patient positioning aids
- Visual inspection.
- Radiolucency.
- Condition.
- Cleaning.

Patient measuring callipers
- Accuracy.
- Damage.

Collimator

The collimator, sometimes referred to as the light beam diaphragm (LBD), provides the radiographer with an easy to use, accurate, method of controlling the size of the X-ray field and placing it over the area of interest, thus reducing the radiation dose to the patient and improving the quality of the image.

The collimator receives much use and is vulnerable to knocks, often resulting in inaccuracy of the light beam/X-ray beam coincidence, blown light bulbs, electrical and mechanical problems.

Visual inspection (see Appendix B, page 142)
- Rotation.
- Stability.
- Knobs for shutter control.
- Accuracy of shutter setting scales (if present).
- Shutter chain drive.
- Shutters.
- Housing.
- Window.
- Light timer.
- Timer switch.
- Cables.

QUALITY ASSURANCE WORKBOOK

- Plugs.
- Light bulb.
- Light/X-ray beam alignment.

Accuracy of scales test

Not everyone uses the aperture size scales on the collimator to set the aperture size, it being more common practice to use visual assessment of the light field size. However if staff do practice this method, it is advisable to make sure that the scales are accurate.

Inaccuracy of scales may lead to inaccurate collimation.

Frequency of test
- 6 monthly.

Equipment required
- Collimator to be tested.
- 100 cm rule.

Method
- Set 100 cm focus to tabletop distance.
- Using the 100 cm FFD (SID) scale on the collimator, set various aperture sizes with the light on and measure the resultant light areas at table top level.
- Compare with settings on the collimator.

Evaluation
- Light area measured should be the same as the setting on collimator.

Action
- If settings are not accurate and the solution is not a simple one, call an X-ray engineer.
- File a report.

Changing a light bulb

A light bulb can fail at anytime without warning. It is important then, to have the correct spare bulb readily available.

Staff should be capable of replacing a faulty bulb. DO NOT ATTEMPT THIS UNLESS YOU HAVE HAD ADEQUATE INSTRUCTION. If unsure, call an electrician.

Frequency
- As necessary.

Equipment required
- Suitable light bulb.
- Cloth or tissue.
- Screwdriver.

Method
- Switch off the power.
- Remove relevant collimator housing.
- Check that spare is the correct type.
- Before removing the old bulb, check to see if the replacement bulb must be fitted a specific way round.
- Replace faulty bulb with the new one.
- If the bulb is of the quartz type, ensure that it is not handled with bare fingers, as body oil on the bulb will shorten its life.
- Replace housing and tighten screws.
- Test.
- Ensure that the spare bulb is replaced as soon as possible.

Light beam/X-ray beam alignment test

The purpose of the light in a collimator is to allow more accurate collimation of the X-ray beam. The light must, therefore, coincide with the X-ray beam. The light beam relies on the accurate positioning of the light bulb and angled mirror inside the collimator.

Should either the mirror or light bulb be dislodged, then errors in collimation could occur, resulting in areas of interest being excluded from the field, or too large an area being irradiated.

Frequency of test
- 6 monthly.
- As necessary.

Equipment required
- One 24 × 30 cm loaded cassette.
- Alignment test tool. Commercially made test tools are available, but a simple alternative is to use eight coins or four paper clips (see Fig 2–1, Fig 2–2).
- Lead marker or ninth coin.

Method
- Make sure that the table is level and the central ray at 90° to the tabletop.
- Place a loaded cassette on the tabletop face up.
- Set a FFD (SID) of 100 cm.
- Switch on the collimator light.
- Centre to the middle of the cassette.
- Collimate to within the edge of the cassette, leaving a 3 cm border all round, that is outside the light field.
- Place the coins in pairs, so that, where the coins touch, coincides with the edge of the light area.
- All four borders of the light must be marked in this way.

MODULE 2. ACCESSORY EQUIPMENT

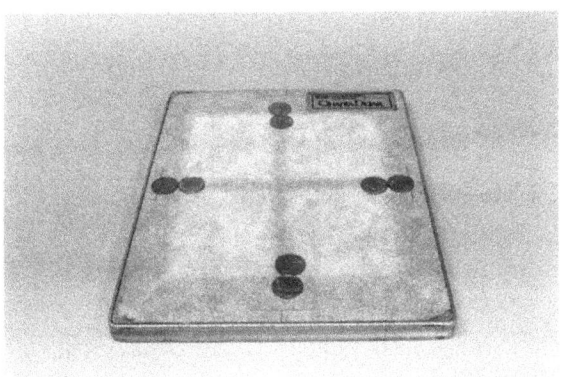

Fig 2–1. Light/X-ray beam alignment test using eight coins

Fig 2–2. Light/X-ray beam alignment test tool. Radiation Measurements Inc.

- Place the lead marker within one corner of the light field so that the film can be related to the light/X-ray field and hence the collimator shutters.
- Make an exposure sufficient to blacken the film.
- Process the film.

Alternative method instead of using coins
- Use four paper clips, each bent to form right angles.
- Place paper clips at corners of light field.

Note: Collimators are known to become less accurate as the field size is increased. To check this make two flash exposures on the same film, following the method described above, using different field sizes, remembering to re position the markers for the second exposure.

Evaluation
- For perfect alignment, the light field (where the coins touch) should coincide with the X-ray field.
- The irradiated area **must *not*** be greater than the area covered by the light.

- At 100 cm FFD (SID) the irradiated area must ***not*** be more than 10 mm smaller than the area covered by the light. This represents a 1% tolerance.

Action
- If the alignment is unacceptable it must be adjusted.
- Call an X-ray engineer.

Note: Cones and diaphragms can also be checked using a similar method.

Shutter efficiency test
Closing the shutters in the collimator fully should prevent any radiation from reaching the film. Useful for testing radiation safety when discharging capacitor discharge mobiles or making tube warm up exposures.

Frequency of test
- 6 monthly.

Equipment required
- One loaded 24 × 30 cm cassette.

Method
- Place the cassette on the tabletop face up.
- Set a FFD (SID) of 100 cm.
- Set an exposure of approximately 80 kV and 40 mAs.
- Open one set of shutters fully, leaving the other closed.
- Make an exposure.
- Fully close the open shutters and open fully the closed ones.
- Make another exposure.
- Process the film.

Evaluation
- Study the film. If the shutters are efficient the film will not have been affected by radiation.

Action
- If the shutters are allowing radiation to pass, call an X-ray engineer.

Cassettes and intensifying screens

Cassettes are the light tight containers that hold the X-ray film between the intensifying screens. They are available in a range of sizes to suit every need.

Intensifying screens fluoresce when struck by radiation, the light emitted significantly contributing to the blackening effect on the film.

Cassettes are easily damaged and likely to wear, resulting in possible light leakage and poor film screen contact.

Intensifying screens deteriorate over time and are easily damaged. Foreign material on their surface or damage will create marks on films.

All cassettes should be clearly numbered on the outside. A corresponding number being placed inside, on the edge of one of the screens, using an indelible marker, where it will not affect the image, but is visible on the radiograph.

All cassettes should be marked on the outside, identifying the type of intensifying screens fitted.

If different film/screen combinations are used in the department, screen speed should be clearly indicated on the outside of the cassette. It is usual for manufacturers to supply appropriate labels with their screens.

Both cassettes and screens should be inspected and cleaned regularly. Records must be kept of all inspections, maintenance and replacements.

Cassettes (see *Appendix B*, pages 143, 144 & 145)

Cassette inspection

Frequency of inspection
- Yearly.
- As necessary.

Equipment required
- Cassettes to be inspected.

Method
- Inspect:
- Hinges.
- Catches.
- Casing.
- Cleanliness.

Evaluation
- Faulty parts, distortion and inadequate cleanliness must be attended to.

Action
- Repair faults or replace cassette.
- Carry out film/screen contact test.
- Clean with damp cloth and wipe dry.
- Keep records of work carried out.
- File report.

Light leakage test

Frequency of test
- Yearly.
- As necessary.

Equipment required
- Cassettes to be tested.

Method
- Load cassette with new film.
- Place cassette under bright light for approximately 15 to 30 minutes.
- Turn cassette and repeat.
- Process film.

Evaluation
- Black fogging around the edge of the film is an indication of light leakage.

Action
- Repair or replace cassette.
- File report.

Note: Similar fogging can be caused by a lid being left off a box of film or the film hopper left open, allowing white light exposure of the upper edge of the films.

Intensifying screens

Most cassettes are fitted with a pair of intensifying screens and should be used with double emulsion film. Blue light emitting screens should be used with blue light sensitive film.

Green light emitting screens should be used with green light sensitive film.

Any damage, deterioration or foreign matter will be seen in resultant images.

Intensifying screen inspection

Frequency of inspection
- Monthly.

Equipment required
- Screens to be inspected.

Method
- Inspect in bright light conditions.
- Screens firmly fitted.
- Correct screens fitted (see label on back of cassette).
- Correct number on edge of screen (see number on back of cassette).
- Condition of screen surface.

Evaluation
- Abrasions.
- Foreign material on surface of screen.
- Discolouration (an indicator of deterioration).

- Fitted correctly.
- Correct screens.

Action
- Loose screens should be re-fitted with double-sided tape.
- Surface damage or deterioration, which is unacceptable—replace screens.
- Clean if necessary.
- File report.

Intensifying screen cleaning

Frequency of cleaning
- Monthly.

Equipment required
- Soft brush.
- Puffer.
- Lint-free cloth, such as gauze.
- Screen cleaner (available from manufacturer) or mild soap (not detergent).

Method
- Clean in bright light conditions.
- Remove all loose dirt with a soft brush or puffer.
- Apply screen cleaner sparingly, with a lint free cloth, such as a gauze swab.
- Use circular motion over whole surface.
- Finish off with long strokes from top to bottom.
- Do not pour cleaner directly on to the screen.
- Wipe with dry lint free cloth.
- Stand cassette open for 30 minutes to dry.
- Inspect.

Evaluation
- Have all dirt particles been removed?
- Are there any smears?

Action
- If dirt particles or smears remain, re clean.
- If, after repeated cleaning, dirt particles remain, decide if they are likely to be a problem.
- If they are considered to be a problem, replace the screens.

Film/screen contact test

If an area of localised blurring is detected on a radiograph, poor film/screen contact should be suspected.

Frequency of test
- Yearly.
- As necessary.

Equipment required
- Cassette to be tested.
- Test tool (box of paper clips or sheet of perforated zinc or fine wire mesh, large enough to cover a 35 × 43 cm film, with a square hole, about 10 cm from one edge, approximately 2 to 2 cm square). (see Fig 2–3 and *Appendix A*, page 134).
- Lead marker if cassettes do not have a lead blanked area for patient details.

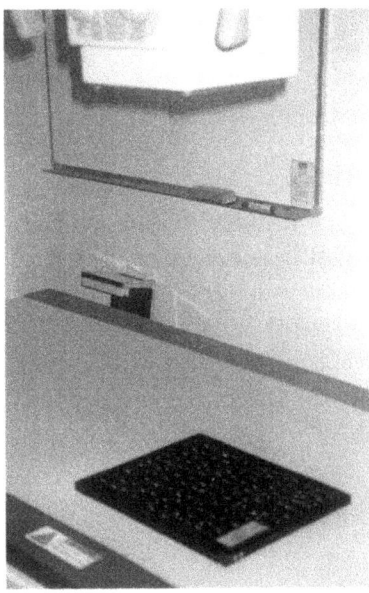

Fig 2–3. Film/screen contact test, using paper clips as a test tool

Method
- Load the cassette to be tested and place it face up on the tabletop.
- Cover the whole of the cassette with the test tool (if using paper clips, distribute evenly).
- Set a FFD (SID) of 150 cm (the longer FFD (SID) reduces geometric unsharpness).
- Collimate to cover whole of cassette.
- If appropriate, place lead marker in the corner of the cassette face, in order to relate resultant radiograph to cassette.
- Make an exposure using about 50 kV and 6 mAs (film density of 1 to 2).
- Process the film.

Evaluation
- If a densitometer is available the film density can be measured at the image created by the hole in the test tool.
- Inspect the image, looking for areas that look blurred.
- A noticeable area of unsharpness could be caused by:

- A damaged cassette.
- Screen packing, deterioration.
- An air pocket.
● When using a close mesh wire test tool the poor film/screen contact areas may also have a higher density.

Action
● Repair or replace cassette.
● Replace packing.
● Re-test.
● File a report.

Film/screen compatibility—colour of light emission test
Different types of intensifying screens may emit different light colours and/or intensities. The colour of light emitted by screens must be the same as the colour sensitivity of the film used with them. Check with your film and screen suppliers if in doubt.

If there is any doubt about the colour compatibility of the light emitted by the screens, the following simple test can be used.

Frequency of test
● As necessary.

Equipment required
● Screens to be tested.

Method
● Open the cassette.
● Remove the film.
● Place the open cassette, screen facing up, on the tabletop.
● Set a FFD (SID) of 100 cm and collimate to cover the open cassette.
● Reduce room lighting to a minimum.
● Make an exposure using a relatively high kV and long time, e.g. 80 kV and maximum time possible.
● *Observe the screens during the exposure. Remember to follow normal radiation safety rules.*

Evaluation
● Note the colour of light given off by the screens.
● Note the intensity of the light given off by the screens.

Action
● If the colour sensitivity of the film is not the same as the screen light emission colour, you should change the type of film you use so that they are compatible.
● If the light intensity comparison between pairs of screens is noticeably different, check the labelling of your cassettes for screen type and/or carry out the **intensifying screen consistency test**.
● Re label the outside of the cassette if necessary.

Intensifying screens consistency test
The efficiency of screens tends to deteriorate over time. If you have screens of the same type, but of differing ages, you cannot assume that they will all give the same result. It is therefore necessary to check the efficiency of all screens periodically in order to achieve consistent results.

Frequency of test
● Yearly.
● As required.

METHOD I

Equipment required
● Step wedge (see *Appendix A*, page 133).
● Loaded cassettes with screens to be tested (all cassettes should be loaded from the same box of film).
● Densitometer, if available.

Method
● Cut a sheet of film into strips large enough to easily cover the step wedge.
● Load each cassette to be tested with one strip of film, placed centrally.
● Place the first cassette on the tabletop, face up.
● Place the step wedge in the centre of the cassette so that it is directly over the film strip inside.
● Set a 100 cm FFD (SID) and collimate to cover the step wedge.
● Set an exposure that will give a full range of densities on the step wedge image (you may need to do test exposures first).
● Expose.
● Process the film.
● Repeat the procedure for all the other cassettes being tested, using exactly the same conditions.
● Label the film strips with date and cassette number.
● For accuracy of this test, the X-ray output and processing must be consistent throughout.

Evaluation
- Place the test strips side by side on the same viewing box.
- Compare the densities (a densitometer will give a more accurate result).
- Assess the density differences between the strips.
- If the film strip images, using screens of similar types, are seen to vary noticeably, this is unacceptable (densitometer readings of similar steps should be within 10% of each other).
- Assess the density differences between the strips.

METHOD 2

Equipment required
- Phantom (water filled flagon, or other suitable, even density material, of at least 30 × 30 cm area). (see *Appendix A*, page 133).
- Loaded cassettes with screens to be tested (all cassettes must be loaded from the same box of film).
- Densitometer, if available.

Method
- Place four of the cassettes together, face up on the tabletop, so that one corner of each cassette meets the others and the appropriate edges are in contact.
- Set a FFD (SID) of 100 cm.
- Centre to the touching corners of the cassettes.
- Place the phantom to partially cover all four cassettes.
- Collimate to within the phantom so that a portion of each cassette will be irradiated.
- Set an exposure that will produce a measurable density on the film and expose.
- All screens to be assessed should be tested in this way, keeping one of the cassettes unchanged throughout. Remember to reload the control cassette each time.
- Identify all films using the cassette numbers.
- Process the films as soon as possible using the same processor.

Evaluation
- Compare the densities produced on each radiograph (a densitometer will give a more accurate result).
- Densitometer readings should be within 10% of each other.

Action
- If significant differences are seen you may wish to re screen all cassettes tested.
- If this is impractical, work out the exposure differences to achieve similar results and label each cassette accordingly.

When equipping a department with all new screens it is advisable to carry out this test as a basis for future comparison.

Grid

The purpose of the grid is to reduce the amount of scattered radiation reaching the film. This greatly improves the quality of the image. Although the quality of the image is improved, it does mean a significant increase in exposure, and therefore dose to the patient.

A grid looks like a simple thin sheet of soft metal, but it is in fact a precision made piece of equipment and easily damaged. It is made up of a large number of fine lead strips interspaced between radiolucent strips. If the lead strips become distorted the grid will be less efficient and irregular densities will be created on the film.

If the grid is used wrongly, abnormal densities (grid cut off) will be produced. e.g. a focussed grid used upside down will create a reduction in density the further the area is from the centre line.

Grid types

Stationary
- A loose grid which can be placed directly over the face of the cassette.
- The grid and cassette size must be the same.
- The grid ratio is usually 6:1 or 8:1.
- Cassettes are available with a built in, or added, grid.

Moving
- A grid used in a potter bucky system, which moves from side to side during the X-ray exposure, in order to defuse out the images of the lead strips.
- The grid ratio is usually 10:1 or 12:1.
- The grid ratio should be 16:1 if used for high kV work.

Parallel line
- Lead strips run parallel to one another in one direction only.
- There are two types of parallel line grid, **focused** and **non-focused**.
- Moving and stationary grids are all of parallel construction.

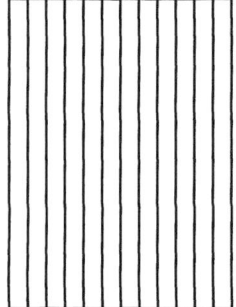

Fig 2–4. Diagram of a parallel line grid. Plan view

Cross-hatch
- Two sets of lead strips superimposed and running at 90° to one another.
- Generally a stationary grid, used only for relatively high kV work and when no tube angulation is necessary.
- This design is used only in stationary grids.

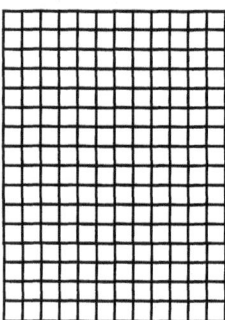

Fig 2–5. Diagram of a cross-hatch grid. Plan view

Non-Focused
- A parallel line grid.
- Relationship of lead strips uniform to one another throughout.

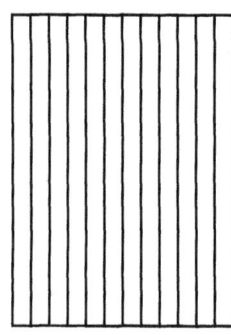
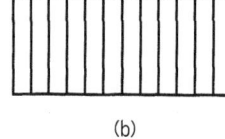

Fig 2–6. Diagram of a parallel line, non-focused grid, (a) Plan view, (b) End view

Focused
- A parallel line grid.
- Differs from the non focused grid only in that the lead strips progressively incline more inward, the further they are from the centre line, so that they are focused to a pre-determined point above the grid.

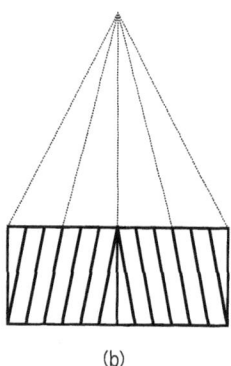

Fig 2–7. Diagram of a parallel line focused grid, (a) Plan view, (b) End view

Structure
- Thin lead strips interspaced between equally thin strips of radiolucent material.
- Covered top and bottom with thin aluminium sheet.

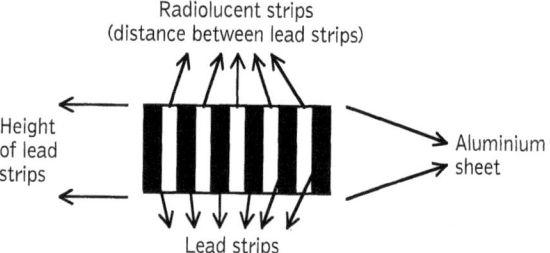

Fig 2–8. Diagram of grid construction, end view

Specifications
The detail of the grid structure should be marked on the surface of the grid either by a label or imprinted into the metal.

MODULE 2. ACCESSORY EQUIPMENT

Grid ratio — The ratio of the **height** of the lead strips to the **distance between them**.

Grid line — The **number of lead strips** in a grid. Stated as the number of lead strips to the centimetre (or to the inch).

Focal range — Grids are designed to be used within **given focus film distances (source image distances)** dependent on the way the lead strips have been focused.

Grid line direction — The direction in which the grid lines travel is indicated by **a line down the centre** of the grid on the tube side. Cross-hatch grids have a second line running at 90° to the first.

Tube side — The tube side is shown by a label stating TUBE SIDE, or an **X-ray tube symbol**. The **lead strips direction line** also indicates the tube side.

Care of the grid

- Stationary grids should have an added, rigid, PVC protective cover.
- When a protective cover is added the grid specifications must be recorded on the new cover for future reference.
- Do not bend, drop or dent the grid.
- Clean regularly.
- Store in a safe place.

Use of a stationary grid

- Use a grid that is the same size as the cassette.
- Place the grid against face of cassette with nothing between it and the cassette face.
- Check that the tube side of the grid is toward the tube.
- Use strips of adhesive tape to hold the grid in place.
- Use the recommended focus film distance (source image distance) for that grid.
- The X-ray beam must not be angled across grid lines.

When using a grid cassette, only the last two points are relevant.

Grid line damage test

Frequency of test
- 6 monthly.
- As necessary.

Equipment required
- Grid to be tested.
- One loaded cassette, same size as grid.
- One lead marker.

METHOD

Visual inspection:
- Dents, bends, creases, damaged corners.
- Specifications marked.
- Cleanliness.

X-ray:
- Place cassette on table top, face up.
- Place grid to cover cassette, tube side up.
- Set a FFD (SID) in the middle of the known focal range of the test grid or 100 cm.
- Centre to middle of grid.
- Central ray at 90° to cassette/grid.
- Collimate to cover cassette/grid.
- Set an exposure of 50 kV 10 mAs (you may need to experiment prior to the test).
- Expose.
- Process film.

Evaluation
- Is grid line pattern regular?
- Is overall density even?

Action
- If the grid line pattern or density is unacceptable repeat the test using FFDs (SIDs) of 10 cm either side of that used in the first test.
- If the images are still unacceptable, replace the grid.
- File a report.

Note: This test can also be used to check the focal range.

Lead rubber aprons and gloves

Lead readily absorbs radiation and is used in many forms and thicknesses in the field of radiation protection. e.g. lead sheet, lead rubber, lead acrylic.

Lead rubber has the advantage of being flexible and can therefore be made into items which can be worn e.g. aprons and gloves. Lead rubber tends to deteriorate over a period of time, with regular use and mishandling. Regular care, cleaning and testing are important (see *Appendix B*, page 146).

Other lead rubber items
- Gonad shields.
- Thyroid shields.
- Small sheets of differing sizes.
- Shielding on screening units.

Care
- Weekly cleaning or as necessary. Clean with soap and water.
- Never fold any item of lead rubber.
- Aprons should be hung up using strong, specially designed hangers.
- Store gloves flat or upright on a custom made holder.
- Do not store near a heat source.

The lead equivalent (effectiveness) of all forms of lead rubber should be known. Aprons for instance are designed only to protect against scatter. They have a lead equivalent of between 0.25 mm and 0.5 mm. Lead sheet may have a lead equivalent of 2.0 mm or 3.0 mm.

Visual inspection test
Frequency of test
- 6 monthly.
- As necessary.

Equipment required
- Items to be tested.

Method
- Check effectiveness and condition of fastenings on aprons (if any).
- Inspect surface covering for tears or sign of deterioration.
- Inspect all seams where appropriate.
- Check flexibility, feeling for cracks not seen on the surface.
- Inspect for cleanliness.
- Check apron hanger devices.
- Assess storage location.

Evaluation
- General condition and cleanliness.
- Possibility of cracks.
- Storage location and hanging devices.
- Are staff using the apron hangers?

Action
- Initiate any repairs or cleaning required.
- Carry out X-ray check.
- Relocate storage location if unsuitable.
- Re-design hangers if unsuitable.
- Re-educate staff if they are not hanging/storing items correctly.
- Replace if necessary.
- File a report.

Test to detect cracking of aprons or gloves
Frequency of test
- Yearly.
- As necessary.

Equipment required
- Screening X-ray unit (if a screening unit is not available, use a general purpose unit and 35 cm × 43 cm cassettes).
- Items to be tested.

Method
- Spread article to be tested on the tabletop.
- Use screening system to examine whole article.
- Use different kVs.
- If a general purpose X-ray unit is to be used, place cassettes under test article and make separate exposures. Ensure that the whole article is examined.

Evaluation
- Look for cracks or other abnormal variations of density in the image.

Action
- Items found to be defective should be replaced.
- Items being withdrawn can be cut up into smaller pieces, avoiding the cracked sections. These can be used for gonad shielding or shielding parts of the film.
- File a report.

Viewing box

A viewing box, or light box, is a simple but important aid to viewing radiographs and is commonly used throughout X-ray departments. It consists of an even light source of sufficient light intensity to allow optimum viewing.

Viewing a radiograph under good viewing conditions is important in order to visualise the maximum amount of information. Regular maintenance and cleaning should be carried out (see *Appendix A*, page 147). All viewing boxes should:

- Be conveniently placed.
- Be safely fixed to a wall or stable mobile structure.
- Be in good working order.
- Be of satisfactory design.
- Have an all over even light.
- Be clean inside and out.
- Be electrically safe.
- Have an adequate light output.

Design
- Two parallel fluorescent tubes of equal light output, with starters.
- A built in, or attached, spotlight is an advantage.
- The film anchor should hold the film firmly.
- A good quality switch.
- Safe electrical wiring.

Cleaning
Dust on the window or tube will reduce the light output. Marks on the window can be misleading.

Frequency of cleaning
- Outside daily.
- Inside 6 monthly.

Equipment required
- Viewing box to be cleaned.
- Clean cloth.
- Screw driver.

Method
- Remove plug from power socket.
- Remove the front window.
- Clean window on both sides (intensifying screen cleaner is a useful cleaning agent as it has anti static properties).
- Clean the back plate and fluorescent tubes.
- Inspect window for damage.
- Make sure that tubes and starters are firmly in place.
- Replace window.
- Test.

Evaluation
- Any dust on inside or outside?
- Any marks on viewing window?

Action
- Wipe over with a cloth.
- Use spirit to clean window of hard-to-remove marks.

Electrical check
This check should be carried out by or under the supervision of an electrician. Switches, fluorescent tubes, starters and electrical connections should be checked.

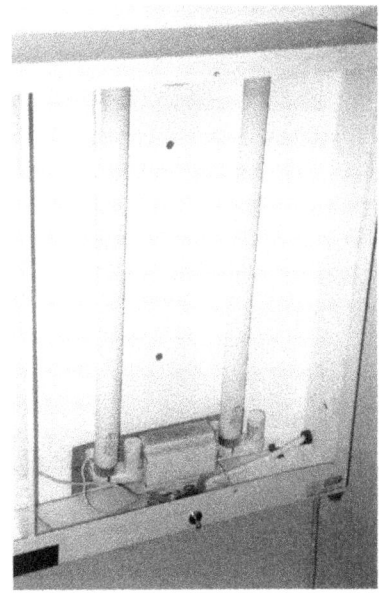

Fig 2–9. Viewing box with front window removed

Frequency of check
- 6 monthly.

Equipment required
- Viewing box to be checked.
- Screw driver.

Method
- Remove plug from socket.
- Remove front window.
- Check condition of wiring.
- Check electrical connections.
- Check stability of switch.
- Check installation of fluorescent tubes and starters.
- Replace front window.
- Replace plug in socket.
- Test.

Evaluation
- Electrical connections firm?
- Electrical wiring in good condition?
- Tubes and starters located properly and working?

- Light source even and comparable to other viewing boxes?
- Replacing fluorescent tubes at different time intervals, or with a different type of tube, may produce a difference in light intensity from one side of the screen to the other or one viewing box to another.

Action
- Tighten electrical connections.
- Refit fluorescent tubes or starters.
- Replace tubes or starters if necessary.
- If, after replacing only one tube, it is found to be brighter than the remaining one, replace both. The working tube that has been removed can be matched with another one at a later date.
- Replace any faulty parts.
- File a report.

Viewing conditions
- Room light levels in the film viewing area should not be too high.
- A bright spotlight should be available for viewing dark areas of the film.
- The viewing box should be mounted at the correct height for easy viewing.
- Light source should be even and the same as other boxes.

Patient positioning aids

Positioning pads

A useful aid to maintaining the position of the patient. A range of pad sizes and shapes should be available in each X-ray room. They must be radiolucent.

Sets of firm foam pads are available from manufacturers (a polyethylene foam is commonly used). Pads may be improvised using other materials but they must be checked for radiolucency.

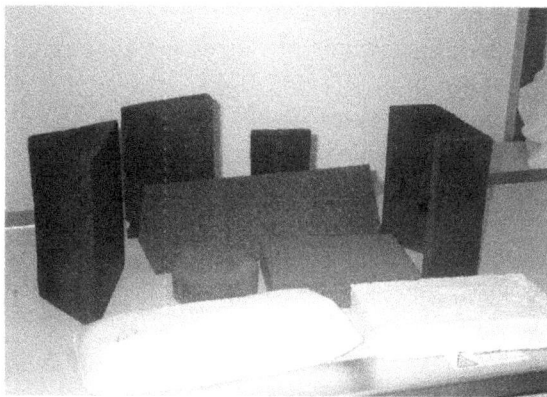

Fig 2–10. Positioning aids—selection of foam pads, sand bag & polystyrene block

Routine care
Care of positioning pads should be ongoing.

- Try to minimise the risk of contamination by:
  — Contrast media
  — Plaster of Paris
  — Blood
  — Urine
- Inspect and wash regularly.
- Do not use for any other purpose.
- It may be preferable to cover with waterproof radiolucent material, or removable cotton covers or both.

Positioning pad tests
- Visual
- X-ray

Visual test

Frequency of test
- Weekly.
- As necessary.

Equipment required
- Pads to be inspected.

Inspect for:
- Likelihood of contamination.
- General cleanliness.
- Shape.
- Smell.
- Particles breaking off.
- Firmness.

Evaluation
- Unhygienic?
- Likelihood of contamination by radio opaque substances?
- Shape, firmness, general deterioration?

Action
- Wash and re-test if necessary.
- If there is a likelihood of contamination with a radiopaque substance then carry out the X-ray test.
- If the pad has deteriorated to the extent that it no longer fulfils its function effectively then replace it.
- File a report.

X-ray test

Frequency of test
- 6 monthly.
- As necessary.

Equipment required
- X-ray unit.
- Pads to be tested.
- Loaded cassettes as necessary.
- Lead marker.

Method
- Place a loaded cassette on the tabletop face up.
- Place the pad on the cassette (ensure that the cassette is big enough to cover whole pad).
- Set a FFD (SID) of 100 cm.
- Collimate to cover the pad.
- Place the marker in one corner of the X-ray field (note which corner of the pad it relates to).
- Set an exposure of 50 kV and 6 mAs.
- Expose.
- Process the film.

Evaluation
- Look for radio opaque shadows within the pad.

Action
- If radiopaque shadows are detected, wash and dry the pad.
- Re-test.
- If opacities are still seen, replace the pad.
- Continued use of the pad may be considered, provided it is identified as containing radiopaque material and not used where it might show on the films.
- File a report.

Other positioning aids

A range of materials can be used to assist in positioning the patient in specific positions, some radiolucent and some not.

Care should always be taken to ensure that radiopaque positioning aids do not obscure the image.

Sandbags	These are **radiopaque**. They should have waterproof or removable cotton or linen covers that must be washed regularly.
Flour bags	Similar to sandbags but are **radiolucent**. They mould to different shapes more readily, but are not so heavy as sandbags.
Compression band	If the band is linen, detach from the end fixing mechanisms and wash as necessary. If the band is plastic, simply wipe over with a damp cloth. The ratchet and fixing mechanisms should be inspected and tested six monthly or as necessary. The band should also be tested every six months, or as necessary, for radiopaque substance contamination.
Water Bottle	Empty fixer or developer bottles full of water are useful as cassette supports, or in weight bearing views of a/c joints. Water is relatively radio-opaque. Little care is required, simply wipe over the outside of the bottle with a damp cloth as necessary. Ensure that the lid is on tightly and that the bottle does not leak.

- Polystyrene blocks used to pack equipment can be cut to required shapes.
- Polystyrene chips used in packing can be sealed into suitable sized cotton or linen bags.
- Cardboard boxes can be cut down to suitable sizes and shapes and re-stuck.
- Wooden blocks can be cut to shape.
- Folded blankets can also be useful.

Note: It should be remembered that any improvised positioning aid must be checked as to its radiolucency before use.

Patient measuring callipers

Measuring callipers are a simple device that allow easy and accurate measurement of a body part.

Calculating radiographic exposure factors can be made more accurate by relating them to the thickness of the body part (see *Module 6. Radiographic exposures*, page 124 and *Appendix A*, page 135). In order to do this, it is more convenient to use measuring callipers.

Care
- Handle with care as they are easily bent or broken.
- Any distortion may mean that they are no longer accurate.
- Store in a safe but convenient place.
- Clean regularly.

Callipers accuracy test

Frequency of test
- Monthly.
- As necessary.

Equipment required
- Callipers to be tested.
- Measuring device.

Method
- Check that the moveable arm is not loose.
- Measure the distance between the outer ends of the two arms at **minimal**, **mid** and **maximum** distances.
- Repeat these measurements at the base of the arms.
- Check against scale.

Evaluation
- The measurements should be accurate to within + or −2 mm.

Action
- If the measurements are not accurate, carefully bend the arm.
- If the arm is loose, attempt to tighten it. If unable to correct faults, arrange to have it repaired or replaced.
- File a report.

Notes

TASK 3
Collimator

Several staff have had to repeat films because they have coned off part of the image. The problem seems to be associated with one X-ray room only. What would you do?

   a) Select an X-ray room and carry out the test you think is appropriate.
   b) Give a brief summary of what you did.
   c) Write a report of your findings.
   d) Recommend appropriate action, if any.
   e) Write your report below.
   f) If there is insufficient space, use additional sheets of paper and attach.

Brief summary of procedure

Report

MODULE 2. ACCESSORY EQUIPMENT

Recommendation for action, if any

_____
_____
_____
_____
_____
_____
_____
_____
_____
_____
_____
_____
_____
_____
_____
_____
_____
_____
_____

Include any test films with this report

Tutor's comments:

Satisfactory/Unsatisfactory

Signed _____ Date _____
     Tutor

TASK 4
Changing a collimator light bulb

The light in the collimator does not work. You suspect a blown bulb.

    a) Select a collimator and carry out the appropriate remedial action.
    b) Answer the following questions in the spaces provided.
    c) Ask your tutor or an electrician to supervise your actions.

1. What type of light bulb in use? _____

2. Does this bulb have to be fitted a certain way round?   YES/NO

3. If YES, describe this: _____

   _____

   _____

4. Where are the spare bulbs kept? _____

   _____

5. Is a correct spare readily available?   YES/NO

6. Who supplies the spare bulbs? _____

   _____

Tutor's comments:

                                                                     Satisfactory/Unsatisfactory

Signed _____   Date _____
           Tutor

MODULE 2. ACCESSORY EQUIPMENT

TASK 5
Film/screens

You have detected a localised area of unsharpness in an image, carry out the test you would do to investigate the problem.

    a) Select any cassette and carry out the appropriate test.
    b) Briefly describe the test you carried out.
    c) Report on your findings.
    d) Recommend any action you consider necessary.
    e) Write your responses in the spaces provided.

Brief description of test carried out

_____
_____
_____
_____

Report on your findings

_____
_____
_____
_____

Recommended action

_____
_____
_____

Include your test film with this report

Tutor's comments:

Satisfactory/Unsatisfactory

_____

Sigature   Date
                Tutor

TASK 6
Intensifying screens

The densities of your radiographs have not been consistent. You are confident that your choice of exposures and the X-ray equipment is not the problem. You suspect the problem may lie with the efficiency of some of your screens.

　　a) Select any three cassettes of similar size, fitted with the same type of screens and using the same type of film.
　　b) Carry out the appropriate test to check the consistency of those screens.
　　c) Briefly describe the test you carried out.
　　d) Inspect the resultant films and make a report on your findings.
　　e) Recommend any action you consider necessary.
　　f) Write your responses below in the spaces provided.

Brief description of test carried out

_____
_____
_____

Report of your findings

_____
_____
_____
_____

Recommended action

_____
_____
_____

Include your test films with this report

Tutor's comments:

Satisfactory/Unsatisfactory

Signed _____   Date _____
　　　　　　　Tutor

MODULE 2. ACCESSORY EQUIPMENT

TASK 7
Stationary grid

It is time to carry out the regular QC checks on your stationary grids.

(a) Select a stationary grid.
(b) Carry out the appropriate stationary grid tests/checks.
(c) Answer the questions below, in the spaces provided.

1. Record the grid specifications:

    Grid Ratio _____
    Grid Line _____
    Focal Range _____

2. Is the tube side identified?   Yes/No

3. How was the tube side identified?
    _____

4. Is the direction of the lead strips indicated?   Yes/No

5. How was the direction of the lead strips indicated?
    _____
    _____

6. Write your visual report here
    _____
    _____
    _____
    _____
    _____

7. Write your evaluation of the X-ray image here
    _____
    _____
    _____
    _____
    _____

QUALITY ASSURANCE WORKBOOK

8. What action do you recommend?

_____
_____
_____
_____
_____

Include your test films with this report

Tutor's comments:

Satisfactory/Unsatisfactory

Signed _____ Date _____
            Tutor

TASK 8
Lead rubber apron and gloves

It is time to carry out the regular QC checks on your lead rubber items.

    a) Select a lead rubber apron and pair of gloves.
    b) Carry out the appropriate tests/checks.
    c) Answer the following in the spaces provided.

1. Write your visual inspection report:
   General Condition:
   _____
   _____

   Cleanliness:
   _____
   _____

   Possibility of Cracks:
   _____
   _____

   Storage Location:
   _____
   _____

   Hanging Devices:
   _____
   _____

   Staff use of Hanging Devices:
   _____
   _____

2. Write your test report:
   _____
   _____
   _____

3. Write your recommendations for action, if any

_____
_____
_____
_____
_____
_____

Include your test films with this report

Tutor's comments:

Satisfactory/Unsatisfactory

Signed _____ Date _____
                 Tutor

TASK 9
Viewing box

It is time to carry out your routine QC checks on your viewing boxes.

a) Select a viewing box.
b) Carry out the appropriate procedures.
c) Ask your tutor or an electrician to supervise this.
d) Answer the following questions in the spaces provided.

1. Is the viewing box
   - Conveniently placed?   YES/NO
   - Safely fixed to the wall/mobile structure?   YES/NO
   - In good working order?   YES/NO
   - Of satisfactory design?   YES/NO

2. Does it have
   - An adequate light output?   YES/NO
   - An all over even light?   YES/NO

What action would you recommend, based on your answers above?

_____
_____
_____
_____
_____
_____
_____

Tutor's comments:

Satisfactory/Unsatisfactory
_____

Signed _____   Date _____
     Tutor

QUALITY ASSURANCE WORKBOOK

TASK 10
Positioning pad inspection

It is time for the regular QC checks of your positioning pads.

    a) Select any one positioning pad.
    b) Carry out the appropriate tests/checks.
    c) Answer the following questions in the spaces provided.

1. Write a visual inspection report
   General Cleanliness
   _____

   Likelihood of radiopaque contamination
   _____

   Shape
   _____

   Firmness
   _____

   Smell
   _____

2. Write your X-ray test report
   _____
   _____
   _____
   _____
   _____

3. Write your recommended action, if any
   _____
   _____
   _____
   _____
   _____
   _____

Include your films with these answers

Tutor's comments:

Satisfactory/Unsatisfactory

Signed _____ Date _____
               Tutor

MODULE 3
X-ray equipment

Aim
To provide the technical information and testing routines that will allow simple care, maintenance and testing to be carried out on basic X-ray equipment. To provide an insight into the choosing and accepting of basic X-ray equipment.

Objectives
On completion of this module the student will have the technical knowledge and practical skills to carry out simple cleaning, maintenance and testing on:

Generator
- Radiation output reproducibility.
- Constancy of radiation output at different mA settings.
- Accuracy of kVp.
- Accuracy of timer.

X-ray tube and high tension (HT) cables
- Electrical.

Tube support
- Mechanical.
- Alignment.
- FFD (SID).

X-ray table
- Mechanical.
- Electrical.
- Alignment.

Tomography
- Mechanical.
- Electrical.
- Cut (layer) level.
- Image quality.

Potter bucky
- Mechanical.
- Electrical.
- Alignment.

Portable and mobile units
- Moving.
- Setting up.
- Mechanical.
- Electrical.
- Checking the X-ray output.

The term "X-ray equipment" covers generator, tube, cables, control panel, table, tube stand and bucky. Specific types of equipment considered are general purpose, tomography, portable and mobile.

It is important to ensure that all equipment is working efficiently and accurate records are kept (see *Appendix A*, pages 133 to 137).

Choosing X-ray equipment

Identify your need
- General purpose, fluoroscopy, specialised, mobile.
- Generator (design/output).
- To suite power supply.
- kV, mA, time ranges.
- Tube mounting (overhead or column).
- Table (tilting, fixed, floating top, bucky).
- Upright bucky.
- Anticipated patient through put.

Resources available
- Power.
- Space.
- Maintenance and repair.
- Finance.

Manufacturer warranty
- Warranty period.
- Warranty cover.
- Determine responsibilities.

Supplier service
- Installation.
- Initial QC tests.
- Maintenance.
- Repairs.
- Parts.
- Speed of response.

Making the choice
- Establish your specification requirements.
- Review manufacturer's specifications.
- Make a short list of suitable equipment.
- Submit your requirements to the selected manufacturers for quotations.
- Review quotations.
- Make selection.

Ensure that you have the written agreement of the manufacturer or supplier, to supply and install the equipment chosen, at the price quoted and that there are no hidden extras.

Acceptance of new X-ray equipment

Upon completion of the installation and before accepting responsibility for the new equipment, you must check that

- Specifications of equipment installed are the same as that ordered.
- Installation has been carried out as stated in the contract.
- Installation is complete and equipment is working efficiently and to your satisfaction.
- All QC tests have been carried out and are satisfactory.
- All accessory equipment has been supplied and is in good order.
- Operating manual has been supplied and is the correct one.

Only when you are sure that the installation has been carried out satisfactorily should you sign the acceptance papers.

Generator

A generator is part of the electrical system of an X-ray machine which determines radiation output. Any inconsistency in radiation output will adversely influence the quality of the radiograph.

The kV, mA and time must be consistent on all settings. Many factors can influence the radiation output. The ones to be considered are:

- Fluctuations in the main electrical supply.
- Inconsistency or inaccuracies in kVp, mA and time.

Fluctuation in main electrical supply
Most X-ray units have a **mains compensator** that automatically compensates for mains fluctuations. However, in some less sophisticated units this must be done by hand. The radiographer must always check the mains compensator meter before making an exposure and make any corrections necessary.

Radiation output reproducibility test
To check that the radiation output is consistent when identical exposures and conditions are used.

Frequency of test
- At the start of a quality control program.
- Yearly.
- As necessary.

Equipment required
- A water phantom, consisting of a plastic flagon or bucket filled to a depth of 10 cm (see *Appendix A*, page 133).
- Two sheets of lead rubber.
- One 24 × 30 cm loaded cassette.

Method
- Place the cassette face up on the X-ray table.
- Cover three quarters of the cassette with lead rubber, crosswise, leaving an end section uncovered.
- Set a 100 cm FFD (SID).
- Collimate the beam to cover the uncovered section.
- Place the water phantom over the area to be irradiated.
- Set an exposure that will produce a low density on the film (**density of 1**). 80 kV and a low mAs is suggested. It should be just possible to read newspaper print through the density produced (you may find it necessary to experiment before starting the test).
- Check that the mains compensator is correctly set.
- Make an exposure.
- Repeat this procedure for each of the remaining three sections. Making sure that you have identical conditions. When placing the lead rubber allow for a space of about 2 mm between each exposed area.
- Process the film.

QUALITY ASSURANCE WORKBOOK

- The same cassette should be used each time the test is carried out.

Evaluation
- Compare the densities. If you have a densitometer, take readings from the centre of each image and compare these too.
- The conditions for all four exposures were identical; therefore all densities should be the same.
- If they are not then the unit is not producing a consistent output.
- Compare with any previous test films.

Action
- If the densities are not the same repeat the test at different times of the day. Mains fluctuations may be the problem.
- If the densities remain inconsistent, call the X-ray engineer.
- If the inconsistencies are minimal continue to use the unit.
- If the inconsistencies are unacceptable, then stop using the unit until the fault has been corrected.
- Mark all films with date and time, and keep for reference.
- Record exposure factors, FFD (SID), water phantom details, size of the X-ray field, cassette number and type of film used, for future reference.
- File a report.

Constancy of radiation output at different mA settings test (see *Appendix B*, page 155)
A test to check the reliability of the mA and time settings (most importantly the mA). The photographic effect for a given **mAs** value should remain constant even though the **mA** and **time** factors may be varied. All other factors being constant.

Frequency of test
- At the start of a quality control program.
- Yearly.
- As necessary.

Equipment required
- A step wedge or 10 cm water phantom, as used in previous test (see Appendix A, page 133).
- Two sheets of lead rubber.
- One 24 × 30 cm cassette.

Method
- Carry out the procedure used for the reproducibility test (see page 34), with the following exceptions.

- Use **three** exposures with the **same** kV and mAs values, but **differing** combinations of mA and time.
- Suggested exposures: (you may find it necessary to experiment before doing the test)
  — Exposure 1: 80 kV 10 mAs (50 mA 0.2 sec)
  — Exposure 2: 80 kV 10 mAs (100 mA 0.1 sec)
  — Exposure 3: 80 kV 10 mAs (200 mA 0.05 sec)

Evaluation
- Despite the fact that the mA and time vary, the **mAs value remains the same**, as does the kV, therefore all **film densities should be the same**.
- If the densities are **not** the same, then it must be assumed that one or more of the exposure factors are inaccurate or inconsistent.
- By varying the mAs and time factors, but retaining the same mA and kV values as before, it should be possible to identify which of the factors is at fault.
- If the out of step density in each test was the second exposure (100 mA), this indicates that the fault lies with the 100 mA setting.
  e.g. 80 kV 50 mAs (50 mA 1.0 sec)
        80 kV 50 mAs (100 mA 0.5 sec)
        80 kV 50 mAs (200 mA 0.25)
- kV and time checks should be carried out in conjunction with this test.

Action
- Repeat the test several times, varying the exposures, but always keeping the kV and mAs constant, within any one test.
- If the densities continue to be inconsistent, call an X-ray technician.
- If the inconsistencies are minimal continue to use the unit.
- If the inconsistencies are large, stop using the unit until the fault has been corrected.
- Mark all films with date, time and exposures and keep for reference.
- Record exposure factors, FFD (SID), water phantom/step wedge details, size of X-ray field, cassette number and type of film used.
- File a report.

Accuracy of timer test
A test to check the accuracy of the X-ray unit timer. This test should only be carried out on a 1 or 2 pulse unit. Timers of 3-phase and capacitor discharge units cannot be checked by this method.

MODULE 3. X-RAY EQUIPMENT

Fig 3–1. Accuracy of timer test using spinning top

Frequency of test
- At the start of a quality control program.
- Yearly.
- As necessary.

Equipment required
- Manually operated spinning top (see Fig 3–1 and *Appendix A*, page 135).
- One loaded 24 × 30 cm cassette.
- Two sheets of lead rubber.

Method
- Place the cassette on the X-ray table face up.
- Place the spinning top in one quarter of the cassette.
- Cover the other three quarters of the cassette with lead rubber.
- Set a 100 cm FFD (SID) and collimate to cover the spinning top.
- Set an exposure of 70 kV, 100 mA and a time of 0.1 sec.
- Manually spin the top. After it has settled to a medium speed, make an exposure (it may be necessary to experiment before starting the test).
- Repeat the process on each of the remaining quarters of the film.
- Process the film.

Evaluation
- The film will show four images, each having a series of spots forming an arc.
- The distance between the spots is not important as this only represents the speed at which the disc rotates.
- The spots should easily be counted, so do not allow the spots to complete a circle and overlap.

- The accuracy of the timer is checked by counting the number of spots.
- **If the mains supply has a frequency of 50 Hz (50 cycles per second):**
  — A one-pulse (self rectified) unit will generate 50 pulses of radiation per second (50 spots on the film).
  — Therefore a **0.1 sec exposure** should produce **5 spots** (see Fig 3–2).

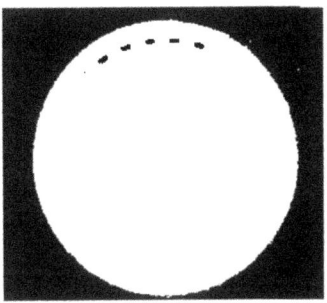

Fig 3–2. Spinning top image of a single pulsed unit using a 50 Hz mains supply

  — A two-pulse (full-wave rectified) unit will generate 100 pulses of radiation in one second (100 spots on the film).
  — Therefore a **0.1 sec exposure** should produce **10 spots** (see Fig 3–3).

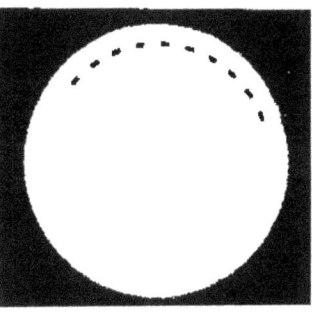

Fig 3–3. Spinning top image of a two pulsed unit using a 50 Hz mains supply

- **If a mains supply has a frequency of 60 Hz (60 cycles per second):**
  — A one-pulse unit will generate 60 pulses of radiation per second (60 spots on the film).
  — A **0.1 sec exposure** will produce **6 spots** (see Fig 3–4).

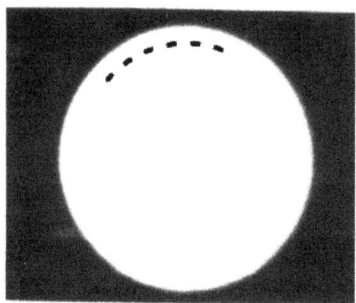

Fig 3–4. Spinning top image of a single pulsed unit using a 60 Hz mains supply

— A two pulse unit will generate 120 pulses of radiation in one second (120 spots on the film).
— A **0.1 sec exposure** will produce **12 spots** (see Fig 3–5).

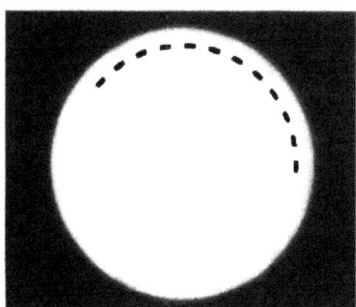

Fig 3–5. Spinning top image of a two pulsed unit using a 60 Hz mains supply

- Other exposure times will produce a different number of spots.

Action
- If the number of spots on the images is **different** from the number expected, the timer may be **inaccurate**. Repeat the test using different time settings.
- If the number of spots on all four images on the film is **the same**, but different from the number expected, the **fault** would seem to be **consistent**. The unit can still be used, but compensatory changes to exposures must be made.
- If the number of spots on the images is **not the same**, the fault would seem to be **inconsistent**. A decision must then be made whether to use the unit or not.
- If the fault appears to be related to a specific time setting, the unit can continue to be used, avoiding this time setting.
- Call an X-ray engineer.
- File a report

Note
- Three-phase and capacitor discharge units will show a **continuous line** image (see Fig 3–6).

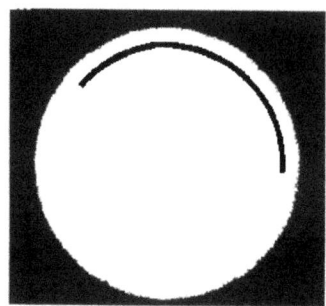

Fig 3–6. Spinning Top Continuous line Image

- These images are of no value in checking the timer.
- The capacitor discharge unit **may** show a continuous line image of **reducing density**.
- If your unit is the type that produces a line of reducing density, you may wish to do spinning top tests on the three highest time settings. Study the low density end of the line. If the density is very low, the radiation at that exposure time is of little value in image production and only adds to patient dose. Consider not using that setting.

Accuracy of kVp test
To check that the generator is producing the kVp set on the control panel. This can be carried out by using a kVp meter or a kVp test cassette. One of the most widely used is the Wisconsin peak kilovoltage test cassette.

Using a kVp meter

Frequency of test
- When the equipment is first installed.
- Yearly.
- As necessary.

Equipment required
- kVp meter.
- Generator to be tested.

Method
- Place the sensor in the middle of the X-ray beam at a FFD (SID) of 100 cm.
- Collimate to a standard area sufficient to cover the sensor.
- Take readings at 10 kVp intervals from 50 kVp to the maximum kVp available.
- Use a standard mA and time throughout.

MODULE 3. X-RAY EQUIPMENT

- Repeat this procedure for each tube focus available.

Evaluation
- Compare each kVp reading with the appropriate kVp setting.
- The measured kVp must be within + or −5 kVp or 5%.

Action
- If the measured kVp s do not fall within the acceptable limits, call an X-ray engineer.

Using a kVp test cassette

Frequency of test
- When equipment is first installed.
- Yearly.
- As necessary.

Equipment required
- Loaded kVp test cassette (see Fig 3–7 (a)).
- Two sheets of lead rubber, one 10 × 23 cm and the other 23 × 23 cm.
- Test cassette calibration graphs for each kVp level used.
- Densitometer.

Method
- Place the cassette on the tabletop, face up (side marked with kVp divisions) and long dimension of the cassette parallel to the anode-cathode axis.
- Set a 100 cm FFD (SID).
- Collimate to the area of the cassette marked with the lowest kVp (60 kVp).
- Set a kVp of 60.
- Set an mAs value that will give a density of approximately 1 in the region of the dots (you may need to experiment).
- Shield the remainder of the cassette with lead rubber.
- Make an exposure.
- Expose each of the other section of the cassette (marked 80, 100 and 120 kVp) separately, using the appropriate kVp each time.
- The time setting should be adjusted for each exposure, to maintain an optical density of 1 in the region of the dots each time. The mA remains the same for each exposure.
- Remember to adjust the lead rubber each time.

Approximate mAs settings

kVp	Single phase unit	Three phase unit
60	500	400
80	75	40
100	15	10
120	12	8

Joel E Gray et al, Quality Control in Diagnostic Imaging.

- Process the film.
- Repeat this process at low, mid and high mAs settings.

Evaluation
- On the radiograph the regions are identified as A (60 kVp), B (80 kVp), C (100 kVp), D (120 kVp).
- Each kVp region contains two columns of dots (see Fig 3–7 (b)).
- The horizontal rows are numbered 1 to 10, 1 being at the darker end of the column.
- One column of dots is of uniform density, in each region, and is called the reference column.
- The dots in the other column, in each region, will show a density gradient, darkest at the top (120 kVp region) and lightest at the bottom (60 kVp region).

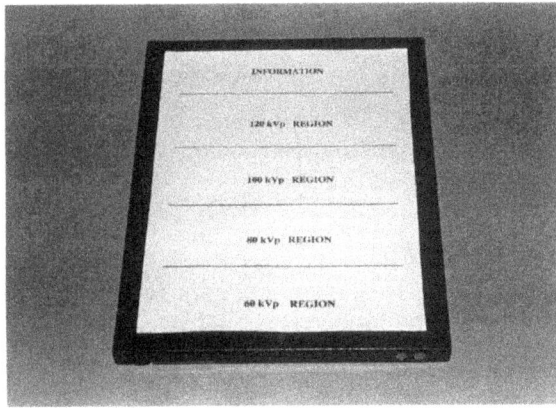

(a)

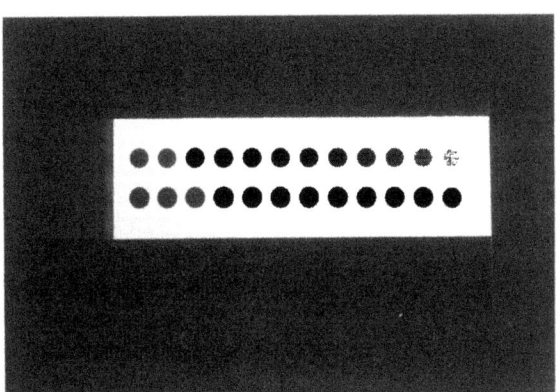

(b)

Fig 3–7. (a) kVp test cassette, (b) test cassette image of one kVp region

- Look for a density match between the two columns, in each region. This should be done with a densitometer.
- If an exact match is not found an average must be determined using the following method.

Example:

Step No.	Optical density of step	Optical density of reference
5	1.40	1.15
6	1.10	1.15

Match step $5 = 6 + \dfrac{1.40 - 1.15}{1.40 - 1.10} = 5 + \dfrac{0.25}{0.30} = 5.38$

- When a match has been determined for each region, refer to the calibration chart for the appropriate region (charts for each of the four kVp regions are supplied with the cassette).
- The chart will have two graphs on it, one for a **single phase** unit and one for a **three phase** unit (see Fig 3–8)

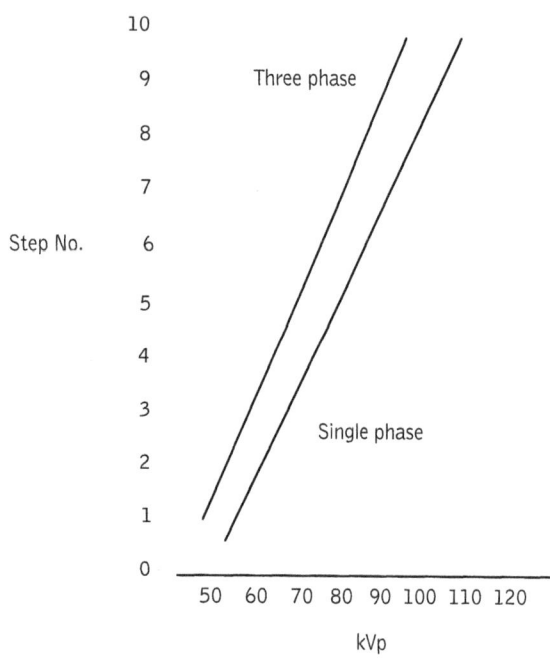

Fig 3–8. kVp test cassette calibration chart, suitable for region C (100 kVp)

- Using the appropriate graph, plot the match step level for each kVp region on the vertical axis, then read off the corresponding kVp s from the horizontal axis. These kVp s should, correspond to the kVp s used.
- **The measured kVp must be within + or 5 kVp or 5%.**

Action
- If the measured kVp s do not fall within the accepted limits, call an X-ray engineer.

Control panel check

Frequency of check
- Yearly.
- As necessary.

Equipment required
- Control panel to be checked.

Method
- Check that all meters are working and all illuminated panels are lit.
- Check that all knobs, switches and buttons are firmly attached and work effectively.
- Check that all read outs are visible and are accurate.
- Check that all signage is readable and correct.
- Check that hand switch operates effectively and is not damaged.
- Check that the cable to any extension hand or foot switch is not damaged or showing signs of wear and that the electrical connections are secure (exposure switches on cables are not recommended).

Action
- Correct any simple faults.
- For more complicated faults call an X-ray engineer.
- Inform staff.
- File a report.

X-ray tube and high voltage cables check

Frequency of check
- Yearly.
- As necessary.

Equipment required
- Equipment to be checked.

Method
- Check that the tube is securely fixed to its support and that any movement locks are working effectively.
- Check that the collimator is securely fixed to the tube and that it rotates freely.
- Inspect the high tension (HT) cables.
  — Check for damage, wear or acute bending.
  — Check that any cable support brackets are

firmly fixed and that the cables lie in them correctly.
— Check that the cables hang freely, do not catch on anything and extend to their required limits without unnecessary tension.

Evaluation
- Anything that is damaged, considered unsafe or has potential for wear or damage should be corrected as soon as possible.

Action
- Correct any simple faults immediately.
- For more complicated faults, call an X-ray engineer.
- Inform staff.
- File a report.

X-ray tube, column, table and upright bucky

Locks and movements check

Frequency of check
- Yearly.
- As necessary.

Equipment required
- Equipment to be checked.

Method
- Test all movement locks for effectiveness and ease of use.
- Test the freedom of movement of the column along the floor and if it travels the full length of the track.
- Any ceiling tracks should be tested in the same way.
- Look for any wear or damage.
- Check for loose or missing screws, bolts or fittings.

Action
- Correct any simple faults immediately.
- For more complex faults, call an X-ray engineer.
- Notify staff.
- File report.

Alignment of X-ray beam to table test

The X-ray beam must be perpendicular to the tabletop, when in the normal over-couch position. Most X-ray tube mountings allow limited angling across the table. The lock for this movement sometimes slips, or is difficult to adjust accurately. This test can also be applied to the upright Bucky.

Frequency of test
- Yearly.
- As necessary.

Equipment required
- Equipment to be tested.

Method
- First check the alignment of the light field and X-ray field in the collimator (see *Module 2, Accessory equipment*, page 30). If an error is found this must be corrected before starting further tests.
- Ensure that the tube is perfectly straight.
- Switch on the collimator light.
- Lower the tube until the window of the collimator is just above the table top.
- Adjust the tube until the centre line of the light field is on the centre line of the table.
- If you have a floating top table, centre the tabletop first.
- Check to see if the automatic centring notch on the tube arm has been located, if there is one.
- Move the tube to the top of the column.
- Look to see if the centre line of the light field has moved away from the centre line of the table.
- Set a FFD (SID) of 100 cm.
- Angle the tube up and down the table, watching to see if the centre line of the light field remains on the centre line of the table.
- Re set the X-ray beam at 90° to the tabletop.
- Move the tube column from end to end of the table, watching to see if the centre line of the light field remains on the centre line of the table.

Evaluation
- If the centre line of the light field does not remain on the centre line of the table during tube movement there is an alignment problem.
- If the centring notch cannot be located when the tube is correctly centred to the table there is an alignment problem. The possible causes are:
  — The X-ray tube has slipped in its mounting.
  — The tube column is not vertical.
  — The cross arm (if there is one) is not horizontal.
  — The tube column track is not parallel to the table.
  — The table is not level.

— The centring notch is wrongly located.
— The tube column track is the wrong distance from the table.

Action
- Carry out the following tests to determine the exact cause of the problem.

Check that column is vertical test

Frequency of test
- Yearly.
- As necessary.

Equipment required
- Spirit level with vertical capability, or a plumb bob (weight on a length of string)
- Equipment to be tested.

Method
- Hold the spirit level vertically against the side of the tube column and note the position of the bubble.
- Move the spirit level a quarter of the way around the column and repeat the procedure, or
- Hold the end of the string high against the centre of the column so that the weight holds the string straight and unobstructed.
- Check to see if the string and column are parallel.
- Move the string a quarter of the way round the column and repeat the procedure.

Evaluation
- If the spirit level bubble is not centred, the column is not vertical.
- If the string is not parallel to the column, the column is not vertical.

Action
- Call an X-ray engineer, to make adjustments.
- File a report.

Check that the table top is horizontal

Frequency of test
- Yearly.
- As necessary.

Equipment required
- A spirit level or an improvised system (bowl or bucket of water) can be used.
- Large diameter bowl or bucket and a ruler.
- Equipment to be checked.

Method
- Place the spirit level horizontally on the centre line, at one end of the table.
- Note the position of the bubble.
- Rotate the spirit level through 90°.
- Note the position of the bubble.
- Repeat this procedure at the other end of the table.
- The bowl of water can be used instead of the spirit level by measuring the height of the water surface above the table around the bowl.

Evaluation
- If the spirit level bubble is not centred, the table is not level.
- If the surface of the water is not parallel to the tabletop, the table is not level.

Action
- If the fault is a problem, it may be possible to put packing under the legs or base of the table or call an X-ray engineer.
- File a report.

Check that the cross-arm is horizontal

Frequency of test
- Yearly.
- As necessary.

Equipment required
- Spirit level or a metre rule.
- Equipment to be checked.

Method
- Place the spirit level on the top of the cross-arm, or hold against the underside of the cross-arm. Note the position of the bubble, or
- Measure the distance from the cross-arm to the table top at the front and back of the cross-arm.

Evaluation
- If the spirit level bubble is not centred, the cross-arm is not horizontal, or
- If the two measurements are not the same, the cross-arm is not horizontal.

Action
- If the misalignment is considered to be a problem, call an X-ray engineer.
- Notify staff.
- File a report.

Test alignment of X-ray beam to upright bucky

- It is usual to align an upright bucky against a wall in line with the end of the X-ray table.
- If this is so, it is useful to have the centre line of the upright bucky in line with the centre line of the table.
- The tube can then easily be angled from the table to the upright bucky.
- Similar alignment tests to those carried out on the table can be used.

Test accuracy of FFD (SID) scale

Many basic fixed X-ray units have a scale fixed to the tube column. This allows the radiographer to easily set the FFD (SID) above the table.

A pointer is fixed to the tube arm mounting, allow accurate measurement to the tabletop. There may be a second pointer giving the FFD (SID) to the under table bucky tray. If there is no second pointer it will be necessary to measure the table top to bucky distance and make allowance when using the bucky.

Frequency of test

- Following new installation
- Following any changes to equipment.
- As necessary.

Equipment required

- Tape measure.
- Equipment to be tested.

Method

- Check to see if there is a pointer and that it is firmly attached and pointing at the correct angle.
- Adjust the pointer if necessary.
- Measure 100 cm from focal spot location dot, on the tube housing, to the tabletop.
- Check to see if the pointer is on the 100 cm mark on the scale.
- If there is a second pointer measure 100 cm from the dot on the tube to the bucky tray.
- Check to see if the correct pointer is on the 100 cm mark on the scale.
- If there is only one pointer, record whether it measures to the tabletop **or** bucky.
- Record the measurement from the tabletop to the bucky tray.
- FFD (SID) readout displays can be checked in the same way.

Evaluation

- Any inaccuracies require remedial action.

Action

- Inform staff of any inaccuracies if the problem cannot be corrected immediately.
- Temporary corrections could be made by sticking a strip of adhesive plaster over the scale and accurately re marking the scale.
- Arrange to have pointers or scale moved to correct location.
- If only one pointer is fitted, arrange for a second to be fitted or notify staff of the adjustment necessary when setting the FFD (SID).
- File a report.
- Carry out the test again after remedial work has been completed.

Tomography

A system which allows prolonged exposures whilst tube and bucky move in harmony.

In simple linear tomography the tube and bucky are connected by an arm, the centre of which is pivoted. The height of the pivot above the tabletop determines the level of image clarity, while all other levels appear blurred.

The size of the angle of swing determines the thickness of cut. The bigger the angle the thinner the cut, the smaller the angle the thicker the cut.

It is a useful technique in blurring out unwanted parts of the X-ray image. It can be a dedicated unit or an attachment to a general-purpose X-ray unit.

Mechanical and electrical check

Frequency of check

- Each time the tomographic equipment is set up if an attachment
- Yearly if a dedicated unit.

Equipment required

- Tomographic unit to be checked.

Method

Inspect:
- Fulcrum attachment attaches securely to the table.
- Connecting arm attaches securely to the bucky and tube.
- Connecting arm pivots correctly.
- Cut height adjustment moves freely.
- Tube mounting column runs freely on tracks.
- Electrical cable, connections and plugs.
- Tomography ON/OFF switch.
- Carry out a test run.

Evaluation
- Electrical cables, switches and plugs fitted properly and show no signs of wear.
- Mechanical parts move freely and show no signs of wear.
- Locks effective.
- Drive motor works effectively.
- Cut height adjustment works effectively.
- No tube vibration during travel.

Action
- Correct any minor faults.
- Call an X-ray engineer for more complex faults.
- File a report.

Cut (layer) height accuracy test

Frequency of test
- Yearly.
- As necessary.

Equipment required
- Tomography test tool (see Fig 3–9, Fig 3–10 and *Appendix A*, page 136).
- One 18 × 24 cm loaded cassette.
- Equipment to be tested.

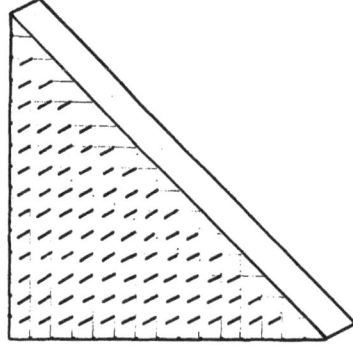

Fig 3–9. Tomography test tool using nails

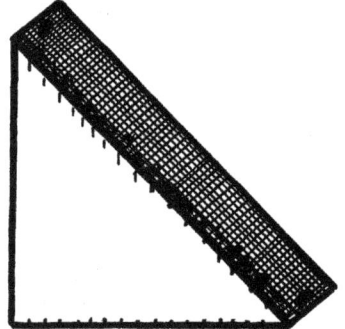

Fig 3–10. Tomography test tool using mesh

Method
- Set up the tomography equipment.
- Place the cassette face up in the bucky.
- Set an exposure of about 50 kV, lowest mA and a normal tomography time.
- Place the phantom longitudinally (nails at 90° to the direction of tube travel) on the centre line of the table, directly over the middle of the cassette when the X-ray beam is vertical.
- Collimate to cover the phantom.
- Set the fulcrum height.
- Place a lead marker within the light field, indicating the height of the fulcrum.
- Position the tube column in its normal start position.
- Make an exposure.
- Observe movements.
- Process the films.

Note: MESH TEST TOOL—must be placed at 45° to the line of tube travel, otherwise there will be incomplete blurring of the wires running in the same direction as that of the tube travel. For ease of use, a wire should be taped across the mesh at the height of the fulcrum. NAIL TEST TOOL—note the number of nails in the row at the height of the fulcrum.

Evaluation
- Study the image. In linear tomography there should be one area of clearly defined, in focus, mesh, or row of nails, with the remainder of the image on either side being blurred.
- Note the height of the sharply defined area in the image and check against the actual cut height. Within + or −5 mm is acceptable.

Action
- You may wish to repeat the procedure with different fulcrum heights.
- If the clearly defined image height and set height do not coincide, carry out the test again to make sure that no errors have occurred.
- If the inaccuracy is still present call the X-ray engineer.
- If the inaccuracy is consistent, the unit can continue to be used, with appropriate temporary changes made to the scale.
- If the observed movements were not smooth, further investigation is necessary.
- File a report.

Cut thickness accuracy test

Although this is not a highly accurate assessment of cut thickness, it is still quite useful. The wire mesh test

MODULE 3. X-RAY EQUIPMENT

tool is more suitable for this test (see *Appendix A*, page 136). A similar test is carried out to that described for the "cut height" test. Repeat the test using different angles of swing, if available.

Evaluation
- Measure the depth of the clearly defined areas of mesh and compare.

Action
- If the cut thickness does not appear to be what it should be, call an X-ray engineer.

Pinhole trace test
The quality of the image can be assessed by using the pinhole trace test

Frequency of test
- Yearly.
- As necessary.

Equipment required
- A sheet of lead, approximately 150 mm × 150 mm, with a 2 to 3 mm diameter hole drilled in the centre.
- One 18 × 24 cm loaded cassette.
- Rule or tape measure.

Method
- With the central ray vertical, place the cassette on the table top, face up, so that the central ray is directed to the middle of the cassette.
- Position the lead sheet at the same height above the table top as the fulcrum, with the pinhole directly over the middle of the cassette.
- Collimate to the hole in the lead.
- Make an exposure with the tube stationary.
- With the same cassette still in position, make a second exposure while the tube moves through its cycle.
- Process the film.

Evaluation
- The radiographic image should be a black line with straight sides and of even density, with a darker spot in the middle of it. The distance from the ends of the black line to the darker central spot should be equal.
- An uneven image indicates vibration of the tube during movement.
- Uneven densities in the image indicate variations in the speed of tube travel.
- Differences in the length of the image either side of the central darker spot indicate that the tube is not travelling the full length of its travel, either at the start or end of its movement.

Action
- Check that the equipment is set up correctly and there is nothing obstructing its travel.
- If the fault is still present call an X-ray engineer.

Note: this test can also be used to check the angle of swing by using the formula

$$\tan \theta = \frac{a}{b}$$

Where a = the distance between the lead sheet and the film

b = the distance between the one end of the image to the central darker spot.

θ = the angle formed by the central ray when vertical and when at the furthest point of its travel during the exposure.

Checking the radiation output
- Use a similar test to that described under Pinhole Trace test (see page 67)
- If the density of the black line is regular throughout (except for the black dot in the middle) the radiation output is considered to be consistent.
- If there is some variation in density then the radiation output has not been consistent.
- This is not an accurate test, but does give some idea of radiation consistency.

Cleanliness of equipment
Cleaning equipment should be treated as part of the quality assurance program. Regular cleaning of all equipment is very important from a hygiene and work environment point of view.

Frequency of cleaning
- Daily
- As necessary.

Equipment required
- Clean cleaning cloths.

Method
- Wipe over all equipment surfaces using a dry cloth.
- If dirt/marks on non electrical equipment cannot be removed, dampen the cloth slightly with water.
- Be very careful when using liquids near any electrical apparatus as most liquids are good conductors of electricity and may cause a short and equipment damage.

- Methylated spirit or cleaning fluids may be used with care, if necessary.

Action
- Keep records of daily cleaning routines.
- Record condition of equipment.

Potter bucky

Mounted under an X-ray table, or in upright form, it consists of a cassette tray, above which is mounted a grid, which move during the exposure. Provides the benefits of a grid, but with the movement diffusing out the image of the gridlines.

Operation test

Frequency of test
- Yearly.
- As necessary.

Equipment required
- Potter-Bucky to be tested.

Method
- Check bucky movement on mounting rails.
- Check efficiency of locks.
- Check bucky tray movement and cassette clamp.
- Check that grid movement takes place during exposure.
- **Remember that this is a test. Close the collimator shutters and direct the X-ray beam well away from the person carrying out the check.**

Action
- If bucky does not run freely on rails:
  — Look for obstruction and free it.
  — Lubricate rails.
  — Request replacements if appropriate.
- If locks are faulty:
  — Adjust or request replacement.
- If bucky tray movement is faulty:
  — Look for obstruction.
  — Lubricate.
- If cassette clamp is faulty:
  — Adjust and/or lubricate.
  — Repair or request replacement.
- If grid movement appears to be erratic, carry out grid movement test.
- If it is not possible to fix the problem, call an X-ray engineer.
- File report.

Grid movement test

The grid in a potter bucky is designed to move from side to side during the exposure, when switched on. Movement may be too fast, too slow, erratic or not take place at all.

Frequency of test
- Yearly.
- As necessary.

Equipment required
- Bucky to be tested.
- One 35 × 43 cm loaded cassette.

Method
- Place the cassette, face up, in the bucky tray and push the tray in.
- Set a 100 cm FFD (SID).
- Centre the X-ray beam to the bucky and collimate to cover the film.
- Set an exposure to give a density of 1 (approximately 55 kV 20 mAs). You may need to experiment before carrying out the test.
- Switch the bucky on.
- Make an exposure.
- Process the film.

Evaluation
- If there is an even, clear pattern of grid lines over the whole of the film the grid did not move at all, or moved too slowly, or for only part of the exposure.
- If there is an uneven pattern of grid lines over the film the grid moved erratically.
- If the grid lines could not be seen, the grid moved, as it should.

Action
- Check to see if anything is stopping the grid movement.
- Check the cable and electrical connections.
- Re test.
- If you have been unsuccessful call an X-ray engineer.
- Inform staff.
- File a report.

Cassette centered to the middle of the bucky test

Frequency of test
- Yearly.
- As necessary.

MODULE 3. X-RAY EQUIPMENT

Equipment required
- Bucky to be tested.
- Marker (small coin).
- One 35 × 43 cm loaded cassette.

Method
- Place the cassette in the bucky tray face up.
- Check that cassette clamp is firmly and correctly closed.
- Push bucky tray fully in and lock bucky in position.
- Set a FFD (SID) of 100 cm.
- Centre the beam to the middle of the table.
- Adjust the bucky so that the central ray coincides with the centreline of the cassette.
- Place the marker on the tabletop so that it coincides with the centre of the X-ray beam.
- Adjust the collimation to cover the cassette.
- Make an exposure.
- Process the film.

Evaluation
- The marker should be in the middle of the film.
- Measure from the image of the coin to all four sides of the film.

Action
If coin image is not in the middle of the film.
- Check bucky tray movement.
- Check cassette clamp.
- Adjust as necessary
- If unsuccessful call an X-ray engineer.
- Inform staff.
- File a report.

Central ray centered to the middle of the bucky test

If the X-ray beam is not aligned correctly to the bucky the grid will create an uneven density image.

Frequency of test
- Yearly.
- As necessary.

Equipment required
- One 24 × 30 cm loaded cassette.
- Test tool (see Fig 3–11 and *Appendix A*, page 137).
- X-ray unit to be tested.

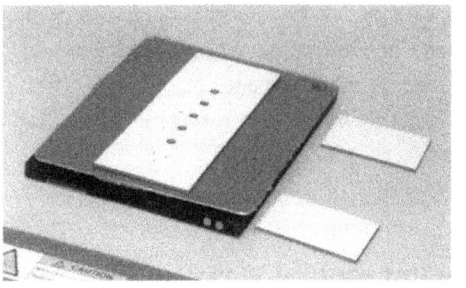

Fig 3–11. Central ray centered to the middle of the bucky test, (a) shielding in place ready for first exposure, (b) test tool in place prior to covering holes

Method
- Switch on the bucky.
- Place the cassette crosswise in the bucky tray.
- Set 100 cm FFD (SID).
- Center the tube to the center of the bucky.
- Place the test tool crosswise on the tabletop so that the central hole (the one with the two small holes either side of it) is directly over the center of the bucky and tape it down. The central ray is therefore centered to the middle of the middle hole of the test tool.
- Note which side the three small identification holes are on.
- Collimate to cover only the center holes.
- Cover all other holes with the lead rubber.
- Make an exposure.
- Do not move test tool.
- Off center the tube so that it is centered over the next hole.
- Adjust the lead rubber so that only this hole is uncovered.
- Leave all other parameters unchanged.
- Make an exposure.
- Repeat the procedure for all five holes and the identification holes, six exposures in all.
- Process the film.

QUALITY ASSURANCE WORKBOOK

Evaluation
- The density of the central hole should be the highest, the holes on either side a little lighter, but of the same density as each other. The two outer holes a little lighter again, but the same density as each other.
- The three small holes are simply to allow correct orientation of the image.

Action
- If the densities of the holes do not appear as described, and tube alignment has been checked, there is some misalignment.
- If this is considered to be a problem call an X-ray engineer.
- File a report.

Portable and mobile X-ray units

Portable

Portable X-ray unit is defined as a small X-ray unit capable of being dismantled and carried from one location to another. It is relatively simple in design, has a maximum kV of about 80–90, a maximum mA of about 30–50, a clockwork timer and a stationary anode tube. Uses a 240-volt power supply.

Moving the unit
- Switch off and remove the plug from the power supply.
- Coil the cable.
- Unplug the component parts and coil cables.
- Dismantle tube stand if necessary.
- The unit may now be carried.
- Take care that you do not drop any of the parts.
- Do not carry too much at the same time.

Setting up the unit
- Place the component parts close enough for the cables to reach.
- Fit plugs firmly into correct sockets.
- Assemble tube stand if necessary.
- Plug in to a 240-volt power supply.

Mobile

Mobile X-ray unit is defined as an X-ray unit on wheels, capable of being moved from one location to another with relative ease. It is bigger than the portable, has a higher output and is usually a little more sophisticated. May or may not have a stationary anode tube. Has an electronic timer. The maximum kV is about 90–100, maximum mA about 50 to 100. May have a pulsed output or capacitor discharge. Uses a 240 volt power supply. More sophisticated mobile units may be medium or high frequency and may have motor driven movement.

Moving the unit
- Mobile units which need to be pushed, have brakes which operate on the "dead man's handle" principle. **The break handle must be depressed to release the brake. Letting go of the brake handle puts the brake on.**
- Units which are motor driven, have a lever or button which operates the motor. When released the motor stops and the brakes are applied.
- Before moving the unit, make sure that the tube arm is in the correct "park" position and firmly locked in place (see operator's manual).
- Remove the power cable plug from the power socket and store in the correct location on the unit.
- Coil the cable neatly and systematically, ensuring that it does not twist and will not tangle when uncoiled.
- If the hand switch is on a cable, make sure that the cable is stowed away so that it will not catch in the wheels or passing objects.
- When pushing the unit ensure that you do not run into anything or any person.
- Take care when passing over uneven ground.
- Take care when going up or down a slope.

Checking the X-ray output
- Similar tests can be carried out to those already described in this module (see pages 67 to 68).

Routine maintenance

Mechanical
- Check all locks.
- Check brakes.
- Check tube arm movements.
- Check collimator (see *Module 2. Accessory equipment*, page 29).

Electrical
- HT cables for wear and abnormal bends.
- Hand switch cable for wear, if fitted (modern safety regulations do not recommend the use of hand switches on extension cables.
- Power supply cable for wear and twisting.
- Plugs and connections.

- Check collimator (see *Module 2. Accessory equipment*, page 29).
- All control panel display lights are working.
- All switches are tight and working.
- Meters working.
- Unit exposes when exposure button depressed.

Any faults must be fixed or reported immediately.

Cleaning
- Dust daily with a clean dry cloth.
- Dirt that resists a dry cloth should be wiped with a cloth very lightly dampened with water (*do not use excessive water as this may get into the electrical components and cause corrosion or shorting*).

Notes

TASK 11
Consistency of radiation output

The density of your radiographs is not consistent, despite using proven exposure factors and techniques.

    a) Carry out an initial test that will demonstrate any inconsistency in radiation output.
    b) Evaluate your results.
    c) Answer the following questions, in the spaces provided.

1. Record the exposure used. _____

2. What is your evaluation of the resultant film?
   _____
   _____
   _____

3. Is this a satisfactory result?   Yes/No

4. What action do you recommend?
   _____
   _____
   _____

Include your test film with these answers.

Tutor's comments:

Satisfactory/Unsatisfactory

Signed _____   Date _____
           Tutor

QUALITY ASSURANCE WORKBOOK

TASK 12

Constancy of radiation output at different mA settings test

You have carried out an initial test and proved that the radiation output is inconsistent. Carry out a test that will attempt to identify the problem.

   a) Carry out a test to investigate the consistency of mA settings.
   b) Evaluate your results.
   c) Answer the following questions, in the spaces provided.

1. Record the exposure used.

   _____
   _____
   _____

2. What is your evaluation of the resultant film?

   _____
   _____
   _____

3. Is this a satisfactory result?   Yes/No

4. What action do you recommend?

   _____
   _____
   _____

Include your test film with these answers.

Tutor's comments:

Satisfactory/Unsatisfactory
_____

Signed _____   Date _____
                Tutor

MODULE 3. X-RAY EQUIPMENT

TASK 13
X-ray timer test

Your radiography densities are still inconsistent. You have already checked the mA and found it to be consistent. Carry out a further test to try and identify the problem.

    a) Carry out a test that will demonstrate inconsistency in the X-ray timer.
    b) Evaluate the results.
    c) Answer the questions below in the spaces provided.

1. Record the frequency of the electrical mains power supply to the unit you used?
   _____

2. Indicate with a tick, which of the following describe the unit you used?
   - One pulse, self rectifies. _____
   - Two pulse, full-wave rectified. _____

3. Is the timer accurate?   YES/NO

4. Explain how you arrived at your answer to question 3.
   _____
   _____
   _____
   _____

5. What action do you recommend?
   _____
   _____
   _____

   Include your film with these answers.

Tutor's comments:

                                                                                                                      Satisfactory/Unsatisfactory

Signed _____   Date _____
                              Tutor

TASK 14
Mechanical and electrical check

The routine mechanical and electrical checks are due on your X-ray unit. Carry out checks for function and safety.

a) Carry out routine mechanical and electrical checks on an X-ray unit.
b) Complete the following check list.

Item to be checked	Yes/No	Comments
Control panel		
Table		
Tube & ht cables		
Tube column & arm		
Other		

Action

Tutor's comments:

Satisfactory/Unsatisfactory

Signed   Tutor   Date

TASK 15
Potter bucky
Electrical & mechanical check

The routine electrical and mechanical check is due on your bucky. Carry out checks for function and safety.

a) Carry out routine electrical and mechanical checks on a bucky.
b) Complete the following check list.

Item	Checked	Comment
Movement—Bucky on rails		
—Tray		
Locks —Bucky		
—Tray		
Electrical —Cables		
—Plugs		
—Switches		
—Grid movement		
Other		

Recommended action

Tutor's report

Satisfactory/Unsatisfactory

Signed _____ Date _____
          Tutor

MODULE 4
Manual film processing

Aim

To provide the knowledge and practical skills necessary to effectively maintain a darkroom and manual processing equipment, carry out manual film processing and establish and run a quality control program.

Objectives

On completion of this module you will be able to understand and carry out manual processing routines, quality control procedures, evaluate and recommend action for:

A darkroom
- Light proof.
- Safe lights.
- Ventilation.
- Storage.
- Condition.

Film and chemical storage
- Stock rotation.
- Conditions.
- Security.

Manual processing
- Processing unit.
- Processing chemicals.
- Processing routine.
- Accessory equipment.
- Health and safety.

The darkroom

The light-tight room in which, the processing of radiographs is carried out. The darkroom is a vital link in the chain leading to standard high quality radiographs. A well positioned and exposed radiograph can easily be ruined by poor processing.

Health issues must also be considered, as the darkroom is invariably a small, confined space where hazardous chemicals are used, under low light conditions (see *Appendix B*, page 156).

The darkroom must:
- Be of suitable size.
- Must not allow white light to enter.
- Have a suitable secure entrance.
- Have appropriate safe lighting.
- Have adequate white lighting.
- Have adequate ventilation.
- Have adequate drainage.
- Have a hot and cold running water supply.
- Have a convenient layout.
- Have a flat smooth surfaced workbench of adequate size and location.
- Be in good condition.
- Be effectively and regularly cleaned.
- Have adequate storage space.
- Have a safe and reliable electrical supply.
- Maintain temperature and humidity at suitable levels.
- Be radiation proof.
- Be fitted out with film processing and accessory equipment.

Other issues involving the darkroom
- Efficient working routines.
- Adequate staff training.
- Regular maintenance checks.
- Suitable fault reporting system.
- Rapid repair and replacement response.
- Safe disposal of exhausted chemistry.
- Safe disposal of contaminated water.
- Safe disposal of empty chemical containers.
- Safe disposal of unwanted film.

Responsibilities of staff using the darkroom
- Use correct processing routines.
- Use the darkroom correctly and safely.
- Carry out correct QC routines regularly.
- Ensure that regular, thorough cleaning is carried out.
- Report or fix all faults immediately.
- Ensure that sufficient stock is maintained.

MODULE 4. MANUAL FILM PROCESSING

White light leaking into darkroom

Film should only be handled in correct safelight conditions. White light should not be allowed to leak into the darkroom.

White light leakage test

Frequency of test
- 6 monthly.
- After work has been carried out on the darkroom.
- As necessary.

Equipment required
- Insulation tape to temporarily cover holes.
- Chalk to mark holes.

Method
- Turn on all lights in areas adjacent to the darkroom.
- Switch off all darkroom lights, including safelights.
- Ensure that any doors are closed.
- Remain in the darkroom for 10 minutes to allow the eyes to get used to the dark.
- Look around the darkroom for signs of white light leaks.
- Pay particular attention to doors, windows, extractor fans, air vent and entry of pipes.

Action
- Identify then seal any white light leaks.
- Repeat the test.
- Carry out a test to see if any film fogging is occurring, if felt necessary.
- File a report.

If you have difficulty in eliminating all white light, e.g. around doors, you may wish to try the following test.

White light fogging test

Frequency of test
- 6 monthly.
- As necessary.

Equipment required
- 1 sheet of new 18 × 24 cm film.
- 1 sheet of 18 × 24 cm card.

Method
- Turn **on** all lights in areas adjacent to darkroom.
- Switch **off** darkroom lights **including safelights**.
- Ensure that any doors are closed.
- Place the sheet of film on the workbench.
- Cover half of the film with the sheet of card.
- Leave for 3 minutes.
- Process the film.

Evaluation
- If the density of the uncovered part of the film is greater than the covered part, then some fogging has occurred.

Action
- Seal any areas of light leakage.
- Repeat the white light leakage test.
- File a report

Safelights

A darkroom should be fitted with appropriate safe lighting. There are several different forms of safelighting available.

- Conventional safelight—small light tight box with a light filter window, fitted with bulb and socket.
- Simple bulb with filter coating.
- Coloured fluorescent light tube.

Safelights should:
- Be in good condition.
- Have the correct filters to suit the light sensitivity of the film used.
- Have the correct light bulb wattage, as recommended by the manufacturer.
- Be electrically safe.
- Be installed correctly.

Check that:
- Each safelight is no less than **130 cm** above the workbench.
- The light bulb is correct (usually 15 watt if facing down).
- The safelight filter is in good condition.
- The safelight filter is compatible with the light sensitivity of the film being used (check with film manufacturer's information).
- The safelight does not leak white light.
- The wiring and fittings are in good condition.

Safelight efficiency test

Frequency of test
- 6 or 12 monthly.

Equipment required
- One 24 × 30 cm cassette loaded with new film.
- Two sheets of 24 × 30 cm card.

- One timing clock or watch with second hand.
- One 24 × 30 cm sheet of lead or lead rubber.

Method
- Place cassette face up on the X-ray table.
- Set a FFD (SID) of 100 cm.
- Cover one third of the cassette with lead rubber, lengthways **(area C)**.
- Collimate to the uncovered area of the cassette.
- Expose the film using a minimum exposure (suggested exposure 45 kV 2 mAs).
- Unload the cassette in the darkroom in total darkness.
- Place the film on the workbench.
- Cover one third of the exposed side of the film with the sheet of card, lengthways **(area A)**.
- Cover areas **B and C** of the film, horizontally, with the second sheet of card, except for a 3 cm strip at the top.
- Switch on the safelights.
- Start clock immediately.
- Wait 30 seconds.
- Move second sheet of card down 3 cm (the first sheet of card must remain in place throughout).
- Wait 30 seconds.
- Repeat this process every 30 seconds until the bottom of the film is reached.
- Switch off the safelights immediately.
- Process the film.

Note: Lightly pre-exposing the film makes it more light sensitive.

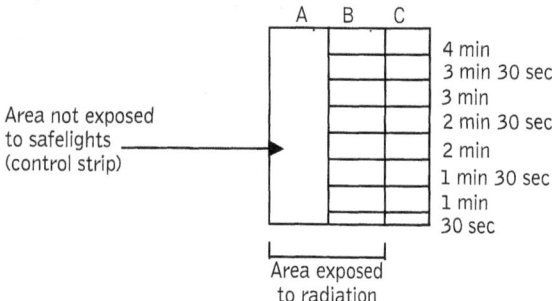

Fig 4–1. Diagram of the test film image

Evaluation (see Fig 4–1)
Section A — Sensitised by radiation.
— *Not* exposed to safelights.
Section B — Sensitised by radiation.
— Exposed to safelights in 8 strips varying in length of exposure from 30 seconds to 4 minutes.
Section C — *Not* sensitised by radiation.

— Exposed to safelight in 8 strips varying in length of exposure from 30 seconds to 4 minutes.
- Identify the strip in Section B which has a noticeable increase in density compared to its equivalent strip in Section C.
- Double check by comparing B with A.
- Note the safelight exposure time of this strip.
- This exposure time is the extreme limit of film handling time.
- 3 minutes is considered to be the limit of acceptable film handling time.

Action
- If the identified safe handling time is considered to be too short, consider one or more of the following modifications:
  — Increase safelight height above workbench.
  — Reduce wattage of bulb.
  — Replace filter.
  — Stop white light leak from safelight.
  — Remove one safelight, if more than one.
- Re-test after modification.
- File a report.

Ventilation

It is important for the darkroom to have adequate ventilation. Ideally, a ducted air conditioning system with light tight vents.

A less efficient, but perfectly adequate system is an extractor fan, ducted to a safe area outside the darkroom. The extractor fan should be sited near the processing unit, with the air intake vent sited at the opposite end of the darkroom to ensure complete air circulation. The air exchange should be at the rate of 15 changes an hour.

Storage

Cassette pass box radiation proof test
If radiation fogging is suspected whilst cassettes are in the pass box.

Frequency of test
- Yearly.
- As necessary.

Equipment required
- 3 coins or lead markers.
- Adhesive tape.
- 1 loaded 35 × 43 cm cassette.

Method
- Stick the 3 markers, at intervals, down the long edge of the front of the cassette with the tape.
- Place the cassette in the pass box, with the markers nearest to the X-ray room.
- Leave the cassette there for a week.
- Process the film.

Evaluation
- If fogging has occurred to the extent that the markers can be seen, repeat the test over a shorter period.
- Establish a period of time when fogging does not occur.

Action
- Do not store cassettes in the pass box.
- Remove cassettes from the pass box within the no fog period.
- Increase lead shielding to pass box.
- File a report

Note: Radiation fogging may be occurring in the X-ray room. Carry out the above test in the cassette storage bin and other areas where cassettes are temporarily or permanently stored.

Accidental light fogging of film

Occasionally staff may leave a film hopper open or the lid off a film box when a white light is switched on. This of course, will fog the films, but not necessarily to the extent that they are unusable.

Due to the fact that films are closely packed, the light may not penetrate much below the top edge.

Degree of light fogging of films assessment test

Equipment required
- Pencil.
- Films to be tested.

Method
- Under safelight conditions:
  — Remove the **front film, rear film** and **one from the middle** of each pack of film that is likely to be affected.
  — Identify which film is which, using pencil marks.
  — Process the films.

Evaluation
- Assess the degree of fogging on each film.
- Decide whether this fogging is acceptable or not.

Action
- If fogging is acceptable or there is no fogging, take no further action.
- If fogging is unacceptable:
  — Use as "clean up" film for an automatic processor, or
  — Discard, or
  — Remove fogged portion by cutting down to smaller size.
- Remind staff to be more careful.
- File a report.

Static electricity

Static electricity charges tend to build up, due to friction. Low humidity and nylon clothing increase the risk of static build up. Most film manufacturers recommend an ideal film handling humidity of 40% to 60%. Static build up on X-ray film is fairly common in dry areas.

Static discharge is seen in the dark as an instantaneous flash of bluish white light. If this discharge takes place on the surface of an unprocessed film, it will be seen as areas of blackening, often "lightning like", on the processed film. If static marks appear on films it will be necessary to carry out remedial work.

Action
- Wipe over all film handling surfaces regularly with an anti-static solution (intensifying screen cleaner may be used as it contains an anti-static agent).
- Clean intensifying screens with a cleaner containing an anti-static agent.
- Instruct staff not to slide films into cassettes or across working surfaces.
- Encourage staff not to wear nylon clothing.
- Check humidity levels. If found to be low, install a humidifier.
- All film handling surfaces, film bins and pass boxes should be electrically grounded.

Film and chemical storage

Film storage

X-ray film should be stored according to manufacturer's recommendations.

Unexposed film is sensitive to radiation, light, temperature, humidity, chemical fumes, bending and pressure. It is important that unexposed film is stored correctly, both in the long term (main film store) and in the short term (darkroom).

All film is given an expiry date, beyond which the fog level is likely to increase. This date can be found printed on the outside of the film box. Some film has been known to age prematurely. Adequate stock rotation is therefore important.

Long term storage
- The store room should:
  — Be well ventilated.
  — Be away from the source of any radiation.
  — Be away from any chemical fumes.
  — Be of adequate size.
  — Have a temperature inside of between **10 °C and 20 °C**.
  — Have a humidity between **40% and 60%**.
  — Be dry.
  — Have shutters or blinds over any windows.
  — Have wooden shelving to avoid condensation.
  — Be kept locked, allowing access only to authorised personnel.
- Store films **upright** to minimise the risk of pressure marks.
- Store films in size groups.
- Do not store films on the floor.
- Use a rotation system so that oldest films are used first.
- Clearly mark all film boxes with the date they arrived in the store to ensure the rotation system will work.
- Film boxes should only be stored one deep, to ensure easy access and correct operation of the rotation system.
- Film boxes should always be **taken** from the **left** and new stock **shelved** on the **right**.
- Keep strict stock records (see Fig 4–2).
  — Film in—Film out.
  — Film in store.
  — Quantities by size.
  — Type of film.
  — Dates.
  — Manufacturer.
  — Supplier.
  — Keep daily record of temperature and humidity.
  — Order stock regularly.
  — Keep record of costs.
- Specialised film with low usage:
  — May need to be kept in a refrigerator to extend expiry dates.
  — Must be in sealed plastic to avoid condensation problems.
  — When removing the film from the refrigerator to use, a period of **8 to 12 hours** must elapse before breaking the seal, to avoid condensation forming on the colder film.

Film stock record

Film size _____ Manufacturer _____

Type of film _____

Boxes of 100 Films

Date	In	Out	Total	Supplier	Destination	Signature

Fig 4–2. Example of a film stock record sheet

Short term storage
- Store only films for immediate use in the darkroom.
- If possible, store films in a specialised, lockable film hopper.
- Preferably a hopper fitted with a micro switch that switches off the white light when opened.
- Place a warning label on the front of the hopper stating "DO NOT OPEN IN WHITE LIGHT".
- Store films by size.
- Standardise the film size location in the hopper.
- Do not overfill the hopper.
- If no hopper is available, store in original boxes, conveniently located.

Care must be taken to ensure that film box lids are always replaced or hopper closed, before switching on the white light or opening the door.

Care must be taken not to damage film boxes.

Chemical storage

X-ray processing chemicals should be stored according to manufacturers recommendations, preferably away from stored films.

Long term storage
- Store in a well ventilated room.
- Store away from the film store. If this is not possible then chemicals must be kept at least **4 metres** away from stored film.
- Room temperature should be between **10 °C and 20 °C**.
- Clearly mark chemical boxes with date of arrival.
- Use a rotation system, **first in first out**.
- **Take** from the **left**, **shelve** on the **right**.
- Do not store too high, this may cause handling problems or injury.
- Keep strict stock records (see Fig 4–3).
- Check regularly for any chemical leaks.
- Clean up any leaks immediately.
- Any windows must have blinds or shutters.
- Keep daily records of temperatures.
- Keep store locked. Allow access only to authorised personnel.

Short term storage
- Only keep a small stock of chemicals in the darkroom.
- Store away from working area.
- Check regularly for leaks.
- Clean up spills or leaks immediately.

Film processing

There are several types of film processing units:

- Manual.
- Bench top automatic.
- Full size automatic—varying degrees of sophistication.

Chemicals stock record

Devel oper/fixer

(Cross out the one not applicable)

Manufacturer _____

Supplier _____

Type _____

Date	In	Out	Total	Supplier	Destination	Signature

Fig 4–3. Example of a chemical stock record

Whichever type of unit is used, a regular quality control program should be followed in order to continually produce high quality radiographs.

Safety precautions in film processing (see *Health and safety,* page 15)

Exposure to processing chemicals and their fumes can be harmful
- Processing chemicals should not come into contact with the skin or eyes.
- Processing chemicals should not be swallowed.
- Processing chemical fumes should not be inhaled.

Wear protective clothing when mixing chemicals:
- Waterproof apron.
- Rubber gloves.
- Goggles.
- Face mask.

First aid

If chemicals come in contact with the skin or eyes:
- Wash thoroughly, immediately. **Emergency eye wash kits should be readily available in all darkrooms.**

Other safety issues
- Chemical spills should be cleaned up immediately.
- Always follow manufacturer's instructions.
- Check processing equipment plumbing for leakage, cracks or blockages.
- Check that all taps/stop cocks work effectively.
- Ensure that processing solution containers are sealed and safely stored.
- Do not consume food, sit next to or use the processor as a source of heat.

- Adequate ventilation (15 air changes per hour is recommended).
- Vent exhaust ducts to an appropriate outside area.
- Check that airflow in the darkroom does remove all fumes adequately.
- Faults and problems should be reported or remedied immediately.
- Safely dispose of exhausted chemistry.
- Safely dispose of contaminated water.
- Safely dispose of empty chemical containers (**do not allow others to use as drinking water containers**).
- Safely dispose of unwanted film.

Notes: Remember that film, film processing chemicals and their containers are hazardous and must not be disposed of in any way that may pollute the environment or cause sickness or injury.

Silver can be reclaimed from exhausted fixer and film. You may be able to sell this to a recognised dealer.

Manual processing

Before setting standards for a quality control program, the processing unit should be emptied, thoroughly cleaned and checked and refilled with fresh chemicals, mixed to manufacturer's instructions.

Manual processing unit design (see Fig 4–4)
There are two common designs for manual processing units:

A common design includes a thermostatically controlled water jacket surrounding all tanks.

A more simple design may have no water jacket, but a thermostatically controlled heater in the developer or no heater at all.

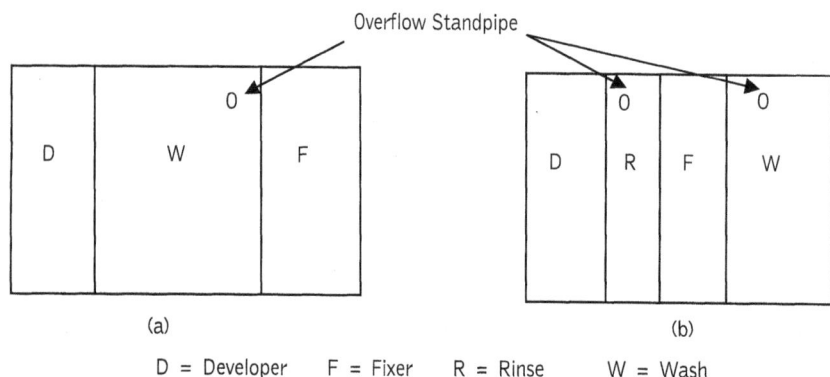

Fig 4–4. Diagrams of the layouts of two different manual processing units

Equipment

- Deep tank processing unit — Developer, fixer, rinse and wash tanks with lids Thermostat. Suitably plumbed.
- Film hangers — Tension type to fit all film sizes. Suitable hanging brackets for hanger storage.
- Drying cabinet — With variable heating control and timer.

  or
- Drying racks — Suspended or wall fixture.
- Thermometer — Alcohol or electronic.
- Hydrometer — Scale range SG 1.075 to 1.115.
- pH paper — Litmus paper.
- Silver estimating paper — Specialised rolls may be available from the film manufacturer.
- Thiosulphate retention test fluid — Bottle of specialised test fluid.
- Timer — Specialised photographic timers are preferred as they are more versatile and reliable.

  — Mechanical kitchen timers are not recommended. Timer accuracy should be checked regularly.
- Chemical stirring rods — 2 PVC or stainless steel rods or paddles. One labelled "**devel oper**" the other "**fixer**". There must be no cross use.
- Measuring cylinder — 100 ml, graduated, glass or plastic.
- Protective clothing — Waterproof apron, rubber gloves, goggles, face mask.
- Cleaning materials — Clean disposable cloths. Clean hand towels. Tank cleaning brushes. Scouring pads (plastic type).

Cleaning

- Switch off any heating or cooling devices in the unit.
- Drain developer, fixer and water tanks.
- Fixer may be retained for silver recovery.
- Ensure that chemistry and contaminated water is disposed of in a safe manner.
- Wash all tanks thoroughly and scrub until clean.
- Use different brush/scouring pad for developer and fixer.
- Flush drainage pipes and ensure they allow free flow.
- Replace overflow stand pipes.
- Close all other tank outlets.
- Scrub inside of film hanger clips.

Mixing chemicals

- Check that tank sizes are standard and that the manufacturer's quantities are correct for your tanks.
- Mix according to manufacturer's instructions.
- Follow manufacturer's safety precautions.
- Use different stirring rods/paddles for developer and fixer.
- Wear protective clothing.
- Check that room ventilation is adequate.
- Minimise splashing.
- Ensure that fixer does not contaminate developer.
- Clean up any spills immediately.

Water supply

- Should have hot and cold running water available.
- Should be reliable.
- Must be clean.
- Do not mix chemistry with water that has a very high chlorine content. This may cause the active fixing agent (sodium thiosulphate) to drop out of suspension, reducing the action of the fixer.

Refilling water tanks

- Refill to top of stand pipes.
- Regulate flow.
- The wash tank water should flow at the rate of four exchanges an hour.

Rate of water flow test

- Time how long it takes for the tank to fill with water once. This should be approximately 15 minutes, to make four changes in one hour.
- Adjust water flow accordingly.
- If too slow, and the tap is fully open, there may be a partial blockage, or this may simply be due to normally low water pressure.

Heating / cooling units

- Switch on.
- Ensure they are working.

QUALITY ASSURANCE WORKBOOK

Clean up
- Wipe over all exposed surfaces.
- Wash and store accessory equipment.
- Safely dispose of all empty chemical bottles and cartons (puncture and place in a sealed plastic bag before disposal).
- Wipe down aprons, gloves and goggles.
- Replace any tank lids.

Before using the unit
- Wait for processing temperature to reach the required level. This may take several hours.
- Check temperature.
- Check water flow rate.
- Check chemical levels.

Developer

The chemical treatment that converts the latent film image into a visual image.

Developer temperature and development time

Development time and temperature directly affect the density, contrast and amount of base fog of a radiograph.

- Follow manufacturer's time/temperature recommendations.
- Use manufacturer's time/temperature adjustment graph when necessary (see Fig 4–5).

Determining the best development time

If local conditions make it difficult to use the manufacturer's recommended development temperature and the time/temperature graph is not available it will be necessary to carry out a test to determine the correct development temperature.

Frequency
- As required.

Equipment required
- Six sensitometry test strips (see *Module 5. Automatic processing*, page 107).
- Thermometer (**not mercury**).
- Six film clips or safety pins.
- Developer stirring rod.
- Timing clock.
- Densitometer.

Method

This test can only provide a development time for one temperature level. If a range of temperatures prevail then this test must be repeated for each temperature required.

- First, regularly check the developer temperature.
- Determine the prevailing temperature.
- Before commencing the test, stir the developer and check the temperature.
- In safelight conditions number the test strips in pencil, 1 to 6.
- Attach the test strips to the film clips (or safety pins) and suspend in the developer. **All strips must enter the developer at the same time.**
- Start the time clock immediately.
- After **30 seconds** remove **test strip 1**, rinse and place in fixer.
- After a **further 30 seconds** (one minute of development) remove **test strip 2**, rinse and place in fixer.
- **Repeat** this process every **30 seconds** until all test strips are in the fixer.
- Fix, wash and dry test strips.

Evaluation
- Arrange strips on a viewing box, side by side, in numerical order, with all unexposed steps at the same end (see Fig 4–6).
- Study the unexposed steps.
- Identify the strips that do **not** show any base + fog level increase.
- Study the middle density steps and identify the strips which do **not** show a density level increase.
- **The most suitable development time, at the temperature used, is the time that gives maximum density but shows no increase in base fog.**

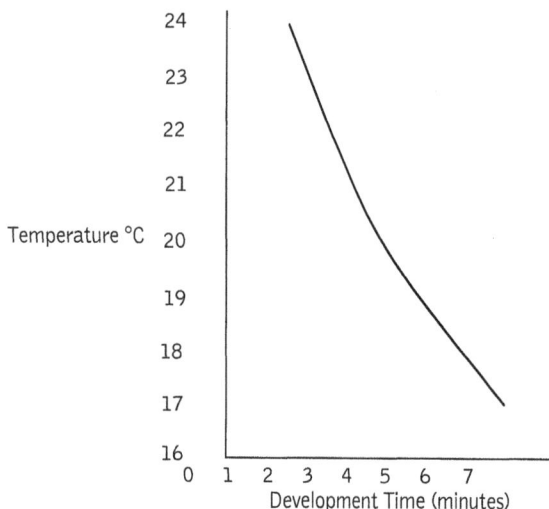

Fig 4–5. Time/Temperature Development Graph

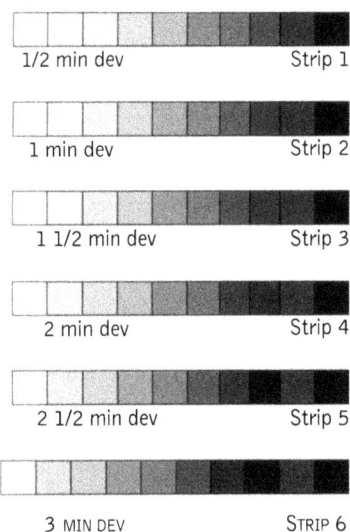

Fig 4–6. Development time assessment strips arranged for comparison

- A densitometer may be used to determine density levels if available.
- If a range of temperatures need to be investigated to determine the best development time for each, the test is repeated for each temperature and the result charted or a graph produced.

Action
- Establish a time/temperature chart (see Fig 4–7).

Emergency measures

To reduce developer temperature
1. ● Fill a strong plastic bag with ice.
   ● Firmly tie off the neck of the bag with cord.
   ● Submerge the bag of ice in the developer or water jacket if there is one.
   ● Agitate.
   ● Constantly monitor developer temperature.
   ● Remove bag when temperature falls to a satisfactory level.
   ● Great care should be taken to avoid bursting the bag in the developer.

   Under no circumstances add ice directly to the developer.

2. ● Use a coil of copper tubing, large enough to just fit into the developer tank.
   ● Fit a hose to one end of the copper tubing, long enough to reach from the tap to the developer tank. Fit hose to tap.
   ● Fit a hose to the other end of the copper tubing; long enough to reach from the developer tank to a drain or sink.
   ● Submerge copper coil in the developer.
   ● Turn on cold water.
   ● Constantly monitor developer temperature.
   ● Remove coil when temperature falls to a satisfactory level.
   ● Turn off water.

To increase developer temperature
- The two methods described above can be used. Replace ice and cold water with hot water.
- Electric immersion heaters can also be use.

Developer activity

The activity of the developer affects the density, contrast and amount of base fog of a radiograph.

It is important, therefore, to regularly check developer activity. Developer activity tests should be carried out immediately after the developer has been renewed and at least once a week.

Some departments do them daily or even twice daily when mammography is involved.

Temperature/development time chart	
Temperature of developer °C	Development time

Fig 4–7. Time/temperature chart used for recording data when assessing development time

Developer activity test (Elementary method)
(see *Module 5. Automatic processing*, page 107, for a more accurate method).

Frequency of test
- Daily.

Equipment required
- One sensitometric strip (see *Module 5. Automatic processing*, page 107).
- Sensitometric strips from previous tests.
- Graph paper.

Method
- Stir developer.
- Check temperature of developer.
- Adjust temperature to pre-determined standard if necessary.
- When temperature is correct, develop the test strip.
- Mark the strip with the date.

Evaluation
- Compare the test strip with the control strip (see Fig 4–9).
- The equivalent steps should have the same density. This confirms consistent developer activity.
- If the strips are not comparable, adjust them until they are.
- If adjustment has been necessary, count the number of steps difference between the two strips.

One step difference   = acceptable
Two steps difference   = acceptable but be warned
Three steps difference = unacceptable take action

- If the test strip is moved **up**, compared to the control strip, the result is regarded as **positive**.
- If the test strip is moved **down**, compared to the control strip, the result is regarded as **negative**.
- Graph the results (see Fig 4–8).

Action
1. If densities are lower than previously:
   - Developer — Exhausted.
     may be — Incorrectly mixed—too much water.
              — Water added instead of replenisher.
2. If densities are higher than previously:
   - Developer — Too active because of
     may be   incorrect mixing.
              — Too much replenishment.
3. Higher basic fog level than previously:
   - Chemical contamination.

Developer activity using a hydrometer test
- An hydrometer will measure the specific gravity (SG) of developer.
- Specific gravity measures the relative weight of a solution, when compared to water, at 23 °C.
- Specific Gravity measurement can therefore be used to show if correct dilution is present.
- The specific gravity (or density) of water is 1.000.

Fig 4–8. Developer activity graph chart

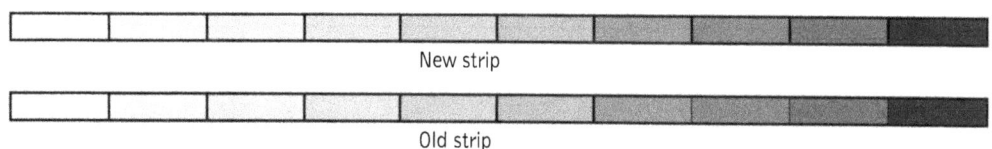

Fig 4–9. Developer activity test strips. Comparison of old against new

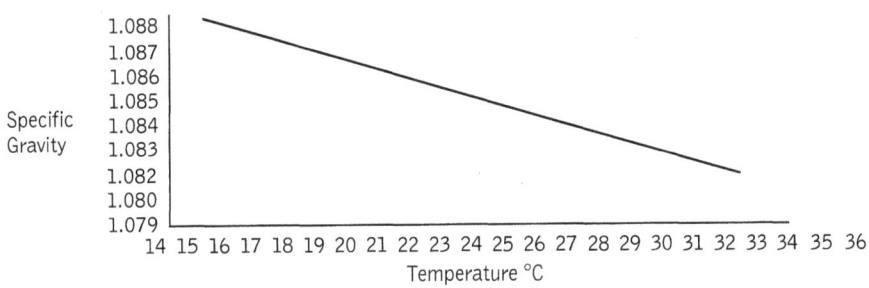

Fig 4–10. Specific gravity/temperature graph

- The specific gravity of developer falls into a range of 1.070 to 1.100.
- Specific gravity will vary with developer temperature (see Fig 4–10).

Frequency of test
- Daily.

Equipment required
- Hydrometer (an hydrometer designed for testing car batteries can be used, but the scale will need to be extended).
- Manufacturer's recommended specific gravity for the developer being tested.

Method
- Place the hydrometer in the developer. It will float.
- Note the reading on the scale at the level of the developer surface.

Evaluation
- The Specific Gravity reading should be within **plus or minus 0.004** of the manufacturer's recommendations.
- If the reading is **lower** than that recommended, the developer is **over diluted**.
- If the reading is **higher** than that recommended, the developer is **too concentrated**.

Action
- If too concentrated add water.
- If too dilute add replenisher or replace the developer.

Developer activity using pH paper (litmus paper) test
Can be used for freshly mixed developer, but is more useful to show contamination or oxidation.

Frequency of test
- Daily.

Equipment required
- pH paper (litmus paper).

Method
- Hold a strip of test paper in the developer for approximately 10 seconds.
- Observe colour change.
- Compare with control strip.

Evaluation
- Note manufacturer's recommended pH range (usually around 10.0 to 10.5).
- Compare test strip reading to recommended range.
- A reading of more than **0.4 below** manufacturer's recommendations **suggests contamination**.
- A reading of **0.8 below suggests oxidation**.

Action
- Replace developer if readings are too high or too low.

Developer replenisher

Replenisher is the chemistry added to the developer to maintain its volume and activity.

- Replenisher is generally more active than the original developer.
- Mix according to manufacturers instructions.
- Have a quantity of replenisher mixed and stored in the darkroom for convenient use.
- Add replenisher when developer level falls.
- Do not allow level of developer to fall below the top of films.

Rinse

The process of using clean water to wash residual developer from the film, so avoiding contamination and reduction in efficiency of the fixer.

- Separate rinse water tanks between developer and fixer should have running water.
- Static rinses should have the water changed regularly.
- Agitate films vertically in static rinses.
- Rinse for 15 seconds.

Fixer

Dissolves off all unwanted film emulsion.
Makes image permanent.
Fixing temperature and fixing time, although important, are not as critical as those of the developer.

Clearing time
- The time it takes to clear the film (dissolving off the unwanted film emulsion).
- Can take from 30 seconds, but should not be more than 2 minutes.
- The film may be viewed in white light once it has fully cleared.

Permanent fixing
- This is usually twice the length of the clearing time.
- It is recommended that all films be fixed for at least 4 minutes to avoid image deterioration later.
- Remove films from the fixer as soon as adequately fixed, as excessive time in the fixer will remove the image.

Fixer activity using a hydrometer test

Specific gravity of the fixer is an indication of its activity (see *Developer activity test*, page 88). The specific gravity of fixer should be in the region of 0.004 (see manufacturer's recommendations).

Frequency of test
- Daily.

Equipment required
- Hydrometer.

Method
- Place the hydrometer in the fixer.
- Note the reading at the surface of the fixer.

Evaluation
- Specific gravity reading should be within **plus or minus 0.004** of manufacturer's recommendations.
- Specific gravity should be in the region of **1.110**.
- A high specific gravity reading—not dilute enough.
- A low specific gravity reading—too dilute.

Action
- Dilute if specific gravity reading too high.
- Add undiluted fixer if specific gravity too low.

Silver estimation test
Indicates level of silver concentration in fixer solution.

Frequency of test
- Daily.

Equipment required
- Silver estimation paper.

Method
- Place paper in fixer.
- Compare with control.
- Take reading.

Evaluation
- Below 2 gm/l is over replenished.
- Above 6 gm/l is under replenishment.

Note: Freshly mixed fixer will have a zero reading.

Action
- Below 2 gm/l add water.
- Above 6 gm/l add undiluted fixer.

Wash

The use of clean water to wash all residual fixer from the film, so avoiding the deterioration of the image over time.

- Wash films for 20 minutes in running water, at a rate of four exchanges an hour, in order to ensure satisfactory long term storage of radiographs.
- Should there be a need to conserve water, then slower flow rates and shorter wash times can be used, but it must be remembered that the washing process will be less efficient and radiographs will deteriorate quicker during storage. Consideration should be given to manual vertical agitation when water flow is slower, or wash times shorter, than recommended, in order to improve the washing efficiency.
- Where there are no facilities for a running water wash, it is recommended that radiographs remain

in the water tank for 2 hours, then given 30 seconds of manual, vertical agitation.

Water level must be above top of film hangers to ensure hangers are completely washed.

Hypo retention test

This test indicates the amount of residual thiosulphate remaining in the film emulsion, after film processing has been completed. An indication of adequate/inadequate washing.

An excessive amount of residual thiosulphate in a film emulsion may cause a brown stain to appear on the film.

Frequency of test
- Weekly.
- As necessary.

Equipment required
- Hypo (thiosulphate) retention test fluid.
- Test strip.
- Radiographs to be tested.

Method
- Put a drop of test fluid on the surface of the film well clear of the image.
- Wait for one or two minutes.

Evaluation
- Inspect the moistened area.
- Compare colour with test strip.

Action
- If colour indicates a high level of thiosulphate, check wash water flow and washing time.
- Adjust and re test.

Drying

To remove all moisture, harden the image and make the radiograph durable.

- Drying cabinet or air drying.
- Ensure that films are fully dried before removing from the hangers.
- For emergency drying, a hair drier on warm can be used:
  — Hold film vertical.
  — Sweep warm air horizontally across surface of film, working from top to bottom.
  — Do both sides.
  — Do not get nozzle of drier too close to film.

Unprocessed film handling

Correct film handling will avoid film damage.
- Handle only under correct safelight conditions, to avoid fogging.
- Handling time should be kept to less than 3 minutes, to avoid fogging.
- Keep hands dry when handling films, to avoid marking the film.
- Hold film only on the edges, with tips of fingers, to avoid marking the film.
- Do not allow film to bend (*bending affects the film emulsion producing artefacts*).
- Do not allow film to slide over any surface (this may create **static marks**).
- Do not allow film to come into contact with any moisture, before entering the developer (water or developer will cause premature development and create dark areas, fixer will stop the development process and create clear areas).
- Tension film correctly in hanger.
- Always handle the film with care.

Labelling film

All films must carry the patient's name, date X-ray examination carried out and any other information considered necessary, to allow correct and easy identification of the radiographs.

Methods of labelling films
- Print directly onto the film with a written or typed ticket and a light source, under safelight conditions in the darkroom.
- Print directly onto the film with written or typed ticket and a light source while the film is still in the cassette. This system can be used under white light conditions (special cassettes and printer are required).
- Write on the film in pencil, under safelight conditions, before processing, and then stick a written or typed label onto the film after processing. Writing on the film in white ink after processing is also acceptable.
- Directly onto the film using radiation. Write or type on a specially prepared radio-opaque slip. This slip is placed on the face of the cassette, within the radiation field, prior to exposure.

The preferred method is to have the patient identifying information in the film emulsion and not written on after processing. The most commonly used is the first method given above.

The printer

An easy and accurate method of radiograph identification.

There are two main types of printer, both using a light source and requiring electricity to power them.

- Darkroom printer for use under safelight conditions only.
- The printer used under white light conditions, requiring special cassettes.

Darkroom printer check

Frequency of check
- 6 monthly.
- As necessary.

Equipment required
- Printing ticket of the type that is routinely used.
- One sheet of 18 × 24 cm unexposed film.
- Printer to be checked.

Method
- Inspect electrical cable and plug.
- Check that the light comes on when activated.
- Check light intensity control is not loose.
- Check that the light is not obstructed.
- Using a pencil, mark the maximum, minimum and two evenly spaced intermediate settings of the light intensity control, around the control knob. Label these marks 1 to 4 starting with the lightest setting.
- Write or type a printing ticket and insert it into the printer using the recommended practice.
- Set the light density control to setting 1 (minimum intensity).
- Under safelight conditions, use a pencil to write identifying numbers 1 to 4 in the respective corners of the film.
- Insert the corner of the film marked 1 into the printer, ensuring that it goes in correctly.
- Activate the light in the printer.
- Remove the film and insert the corner marked 2.
- Re set the light intensity control to setting 2.
- Activate the light in the printer.
- Repeat this procedure using setting 3 for corner 3 and setting 4 for corner 4.
- Process the film.

Evaluation
- Electrical cable wear or damage?
- Plug damage or unsafe electrical connections?
- Is anything obstructing the light?
- Does the light work?
- Is the density control knob tight?
- Study the resultant test film.
- Do the printed areas get progressively darker in the right order (1 to 4)?
- Which of the printed areas is the most satisfactory?
- Is the clarity of this image acceptable?
- Is this the setting that is routinely used?

Action
- If the printer is unsafe or is not working efficiently arrange for repair.
- Select the image that best demonstrates the typed/written information.
- If this is the setting routinely used take no action.
- If the selected setting is different from that routinely used, inform staff of the new setting.
- If the writing/ typing is not clearly seen in any of the images, carry out the test again using a different thickness of ticket and/or looking for ticket movement when printing, until a satisfactory result is obtained.
- File a report.

Processing routine

The method by which films are processed in order to produce high quality images.

Method
- Stir solutions.
- Take the developer temperature. Read while thermometer is still in developer.
- Check correct development time at this temperature. Refer to manufacturer's time/temperature graph or your own time/temperature chart (see page 87).
- In safelight conditions load film into hanger.
- Keep fingers to edge of film only.
- Ensure correct tension of film in hanger.
- Set timer for correct developing time.
- Lower hanger and film into developer.
- Hold hanger so that fingers do not come into contact with developer.
- Agitate film in vertical direction two or three times to remove air bubbles and distribute developer evenly over the film.
- Place hanger in developer so that film is fully immersed.
- Agitate film every 30 seconds during development.
- When timer alarm sounds, lift film from the developer and **drain into the rinse** (exhausted

developer should not be drained back into the developer).
- Lower film into rinse for 15 seconds (agitate if rinse is static).
- Lift films from rinse and allow to drain back into the rinse **(rinse water should not be drained into the fixer as this will result in dilution of the fixer)**.
- Lower film into the fixer.
- Agitate two or three times.
- Ensure that film is fully covered by fixer.
- The film may be viewed in white light conditions once it has cleared.
- Leave film in fixer for at least 4 minutes to fully fix.
- Lift film from fixer and transfer directly to wash (exhausted fixer from the film should not be allowed to drain back into the fixer tank).
- Leave in running water for 20 minutes.
- Lift film from wash and allow to **drain back into the wash tank**. This will minimise amount of water being transferred to drying area.
- Place film in drying cabinet or on an air drying rack.
- **Do not remove until fully dry.**

Notes
- All film movement in the chemicals or water must be carried out in a vertical direction or the film may come off the hanger.
- When moving or draining films, always tilt away from you, to avoid chemicals or water draining back toward you.

Notes

TASK 16
Assess a darkroom for white light leaks

An increasing number of radiographs have a grey fog. Equipment, radiation safety measures and processing have all been checked and found to be satisfactory. Check to see if the problem lies with the darkroom itself.

   a) Select a darkroom.
   b) Carry out the white light leak test.
   c) Answer the following questions in the spaces provided below.

1. Were there any white light leaks?   YES/NO

2. Briefly describe where you saw light: _____

_____

_____

_____

3. Are these white light leaks likely to be a problem? _____

_____

_____

4. What is your recommended action? _____

_____

_____

_____

Tutor's comments:

Satisfactory/Unsatisfactory

Signed _____   Date _____
              Tutor

TASK 17

Are films being fogged by white light leaking into a darkroom?

You have noticed white light leaking into your darkroom around the door. It will be difficult to fix it. Check to see if this light is likely to fog films.

    a) Select a darkroom.
    b) Carry out the test to see if white light leaks are fogging films.
    c) Evaluate the film.
    d) Answer the following questions in the spaces provided below.

1. Outline the procedure you carried out: _____

2. Give your evaluation of the film: _____

3. What action do you recommend based on your evaluation of the film: _____

Include your films with these answers

Tutor's comments:

Satisfactory/Unsatisfactory

Signed   Date
      Tutor

TASK 18
Safelight evaluation

Film is being slightly fogged. You have checked the darkroom for white light leaks, the cassettes are in good condition and it isn't radiation fog. You suspect the safelight.

   a) Carry out a safelight evaluation test.
   b) Evaluate your results.
   c) Answer the following questions in the spaces provided below.

1. How many safelights are there? _____
2. What colour are the safelight filters? _____
3. What is the colour sensitivity of the films used? _____
4. Is the safelight filter colour the correct one? _____
5. What condition are the filters in? _____
6. What is the wattage of the light bulb in the safelights? _____
7. Is this the correct wattage?   YES/NO
8. What is the distance between the safelights and the workbench top? _____
9. Is this distance acceptable?   YES/NO
10. If the distance is not acceptable, why not? _____
11. Are there any white light leaks from the safelights? _____
12. What is your evaluation of the test film? _____

_____
_____
_____

13. Based on your answers, what action do you recommend?

_____
_____
_____
_____

Include your test films with these answers.

Tutor's comments:

Satisfactory/Unsatisfactory

Signed   Date

Tutor

MODULE 4. MANUAL FILM PROCESSING

TASK 19
Film and chemical stores

You are concerned about the conditions under which your films and processing chemicals are stored.

    a) Assess the long term storage of films and chemicals.
    b) Answer the following questions in the spaces provided.

1. Are films and chemicals stored separately?   YES/NO
2. Are the locations of these stores satisfactory?   YES/NO
3. If not, why not? _____

_____

_____

4. How are the films stored? _____

_____

5. Is this satisfactory?   YES/NO
6. If not, why not? _____

_____

_____

7. How is satisfactory stock rotation achieved?

    Films: _____

_____

_____

    Chemicals: _____

_____

_____

8. What is the temperature inside the Film Store? _____
9. Is this temperature satisfactory?   YES/NO
10. If not, why not? _____

_____

QUALITY ASSURANCE WORKBOOK

11. Are the film and chemical stores kept locked?   YES/NO
12. What records are kept regarding stock movement? _____
_____
_____
_____
_____
_____

13. Based on your evaluation of the film and chemical stores, what recommendations would you make?
_____
_____
_____
_____
_____
_____
_____

Tutor's comments:

Satisfactory/Unsatisfactory

Signed _____   Date _____
            Tutor

TASK 20
Safety in film processing

You have just been appointed safety officer for your department. You need to find out if the darkroom is a safe working environment.

- a) Select a darkroom.
- b) Make a check list and investigate the points on it.
- c) Make brief comments on your findings alongside each point on your list.
- d) In each case write in your recommended actions.

Write your findings below.
If there is insufficient space, use additional sheets of paper and attach.

Tutor's comments:

Satisfactory/Unsatisfactory

Signed

Tutor

Date

MODULE 4. MANUAL FILM PROCESSING

TASK 21
Checking manual development time

Radiographic density is inconsistent, you have fully tested your X-ray unit, the chemistry is freshly mixed and the developer thermostat is working effectively. You suspect that the development time is not appropriate for the prevailing conditions.

   a) Select a manual darkroom.
   b) Carry out a manual development time check.
   c) Answer the following questions in the spaces provided.

1. What is the commonly used development time? _____

2. How is this time arrived at? _____

3. What is the developer temperature after stirring? _____

4. What is your recommended development time after evaluating the test strips?

   _____

5. How does this temperature compare to the established practice for this darkroom?

   _____

   _____

6. What is your recommendation for action? _____

   _____

   _____

Include your test films with these answers.

Tutor's comments:

Satisfactory/Unsatisfactory

Signed _____ Date _____
       Tutor

QUALITY ASSURANCE WORKBOOK

TASK 22
Assess developer activity

Radiographs consistently look light and lacking in detail. You have checked development time and temperature.

    a) Carry out a developer activity check.
    b) Carry out the same test 2 days later with the same processing unit.
    c) Answer the following questions in the paces provided.

1. Developer temperature at first test: _____

2. Developer temperature at second test: _____

3. What is your evaluation of the test strips? _____

4. What are your conclusions? _____

5. What are your recommendations for action? _____

Include your test films with these answers.

Tutor's comments:

Satisfactory/Unsatisfactory

Signed _____ Date _____
        Tutor

TASK 23
Manual film processing

Radiographic quality has fallen. It is NOT related to positioning, exposures or X-ray equipment. Someone has suggested that you are getting lazy with your manual processing. Check out your processing routine. Answer the following questions in the spaces provided.

1. What was the development time? _____

2. How did you select the development time?
   _____
   _____

3. What was the film clearing time? _____

4. What was the full fixing time? _____

5. Why should the film be drained into the rinse after development? _____
   _____
   _____

6. Why should the film be drained into the rinse after rinsing? _____

7. Why should the film be drained into the wash after fixing? _____
   _____
   _____

8. Why agitate the film? _____
   _____
   _____

9. Why agitate the film vertically? _____
   _____
   _____

QUALITY ASSURANCE WORKBOOK

Tutor's comments:

Satisfactory/Unsatisfactory

Signed   Date

Tutor

MODULE 5
Automatic film processing

Aim

To provide the knowledge and practical skills necessary to use an automatic processor correctly and establish an effective, regular maintenance and quality control program. To provide an insight into choosing and accepting a new automatic processor.

Objectives

On completion of this module you will:

- Have an understanding of the function of automatic processors.
- Have the knowledge and skills to carry out basic maintenance checks on automatic processors.
- Have the knowledge and skills to set up a quality control program for automatic processing.
- Have the knowledge and skills to carry out specific quality control tests on automatic processors.
- Have the knowledge to be able to evaluate the quality control results and recommend remedial action.
- Have an understanding of how to choose a new automatic processor and the acceptance procedures following installation.

Automatic film processing follows the same basic principles as those of manual processing, but under automated and controlled conditions. The automatic processor and its chemistry must be maintained at a high level of efficiency. Regular quality control monitoring and maintenance is essential to ensure that high quality images can be produced consistently.

There is a range of automatic processors available, from the small, simple bench top to the deep tank, more complex, models. The principles and methods used to monitor processor performance are the same regardless of the type of automatic processor used.

Choosing an automatic processor

Identify your needs
- Throughput of films.
- Complexity of unit.

Resources available
- Power.
- Water.
- Space.
- Exhaust systems and ventilation.
- Maintenance and repairs.
- Finance.

Manufacturer warranty
- Warranty period.
- Warranty cover.
- Determine responsibilities.

Supplier service
- Installation.
- Initial QC tests.
- Maintenance.
- Repairs.
- Parts.
- Chemistry.
- Speed of response.

Making the choice
- Establish your processor specification requirements.
- Do you want a silver recovery unit?
- Review the manufacturer's specifications.
- Make a short list of suitable processors.
- Submit your requirements to the selected manufacturers for quotations.
- Review quotations.
- Make your selection.

Acceptance of new processor

Upon completion of the installation and before accepting responsibility for the new processor, you must check that:

- Specifications of the processor installed are the same as that ordered.
- Installation has been carried out as stated in the contract.
- Installation is complete and equipment is working efficiently and to your satisfaction.
- All QC tests have been carried out and are satisfactory.
- All accessory equipment has been supplied and is in good order.
- Operating manual has been supplied and is the correct one.

Only when you are sure that the installation has been carried out satisfactorily should you accept the processor.

Setting up an automatic processor

The manufacturer or supplier should:

- Ensure that pre installation requirements are adequate, e.g. ventilation, water pressure, drainage, power supply.
- Carry out the installation correctly and make sure it is working properly.
- Carry out the initial quality control procedures and leave the results for future reference.
- Give staff adequate instruction regarding its use, care and maintenance.
- Provide the relevant operating manual.
- Provide a contact for service and repairs.
- Provide service under the conditions of the warranty.
- Provide ongoing maintenance if required by the buyer.

Principles of the automatic processor

- The film is transported through the processor by motor driven rollers, at a standard, regulated speed.
- The film passes through the developer, fixer, wash and drying sections for pre set periods of time.
- The developer and fixer are automatically replenished, by pre determined amounts, each time a film is fed into the processor.
- The replenishment rates can be adjusted to cater for the size and quantity of films being put through.
- The wash water, either runs continuously, or only when films are fed through, depending on processor design.
- The temperature of the chemistry is maintained by thermostatically controlled heating elements.
- Film drying is carried out by, variable temperature, hot air blowers.

Use of an automatic processor

- Follow manufacturer's instructions.
- Place the film on the feed tray.
- Ensure that the longer edge of the film leads and the shorter edge is in contact with the side of the feed tray **(smaller films may not make contact with the replenishment microswitches)**.
- Advance the film until it is taken up by the first set of rollers.
- Do not advance the film until the audio and/or visual signal indicates that the previous film has advanced far enough.
- Processing time, dry to dry, could be 60 seconds to 4 minutes, depending on the type or make. Most manufacturers make a range of processors to suite all needs, with varying degrees of sophistication and processing times.

Potential problems

It is important to be aware of potential problems in order to watch out for them during use and inspections.

Mechanical
- Broken cog teeth.
- Roller wear or splitting.
- Transport system breakdown.
- Film jam.
- Film damage.

Electrical
- Power failure.
- Drive motor failure.
- Heating element failure.
- Drying section failure.
- Replenishment pump failure.
- Recirculation pump failure.

Chemistry
- Wrongly mixed.
- Contaminated.
- Inadequate or incorrect replenishment.
- Blocked or partially blocked replenishment hoses.
- Incorrect temperature.

Water
- Inadequate supply.
- Blocked or restricted flow.
- Contaminated.
- Algae growth.

Cleanliness (lack of)
- Marked films.
- Contamination of chemistry.
- Build up of chemical deposit resulting in mechanical failure.
- Health hazard.

Processor maintenance schedule

A processor maintenance schedule should be followed regularly in order to detect or forestall faults. Read the maintenance and cleaning section of the operator's manual. Closely follow the manufacturer's recommendations. Remember that cleanliness is vitally important.

Following are suggested schedules you may find useful

DAILY

Before start up:
- Remove crossovers and wash in warm water, with a sponge or plastic cleaning pad **(always do developer first, then fixer, to avoid contamination of developer)**.
- Wash tank covers and splash guards.
- Wipe over all deep rack rollers that are above solution levels.
- Clean all interior exposed surfaces.
- Check replenishment tanks/bottles levels, color and smell.
- Check replenishment hoses for bends or leaks.
- Replace water drain stand pipe if appropriate.
- Turn on water and check that the tank is filling (time flow if felt necessary).
- Replace crossovers and tank lids.

Start up
- Switch processor on (you will need to activate microswitches manually with the lid off).
- Listen for any abnormal noise or vibration.
- Check film transport system.
- Check replenishment system is working.
- Replace processor lid.
- Feed in one unprocessed 35 × 43 cm film **(do not use processed film as these are harder and contain fixer)**.
- Inspect processed "clean up film". Feed in a second film if felt necessary.
- Clean exterior surfaces, including the feed tray and receiving bin. **Pay particular attention to the feed tray.**
- Wipe over all darkroom work surfaces.

- When the processor has reached normal operating conditions carry out routine sensitometry procedures (see *Sensitometry*, page 112).

Normal working
- Follow operating instructions.
- Constantly be aware of any abnormal noises, changes in operation, leaks or deterioration of processed films.
- Do not pull processed films out before they are clear of the rollers.
- Do not allow anyone to stand next to or lean against the processor.

Shut down
- Remove processor lid. **Remember** it will be necessary to reactivate microswitches when lid is removed (see manufacturer's instructions).
- Observe level of solutions.
- Listen for abnormal noise or vibration.
- Observe transport system.
- Look for any leaks.
- Switch processor off.
- Remove and wash all crossovers, splash guards and tank lids.
- Wipe over all deep rack rollers above solution levels.
- Inspect and wash roller-drive cogs and drive mechanisms.
- Replace tank lids.
- Turn off wash water if appropriate.
- Remove water drain stand pipe if appropriate.
- Wash off all chemical splashes on interior exposed surfaces.
- Wipe any splashes from exterior surfaces.
- Replace processor lid, leaving it slightly raised at one end to avoid build up of fumes and condensation.
- Door of darkroom should be left open with ventilation system operating, if power supply constraints allow.
- Place crossovers on top of processor, with wash stand pipe if appropriate, and cover with a cloth, or store in a convenient cupboard.
- Re stock, chemicals, films and any other depleted supplies.
- Record all restocking.
- Report all faults.

WEEKLY
- **Follow manufacturer's recommendations.**
- Check solution temperatures, in particular the developer (see *Temperature check*, page 110).

QUALITY ASSURANCE WORKBOOK

- — Compare with any readout on unit and manufacturers recommendations.
- — Adjust if necessary.
- Check replenishment rates (see *Replenishment rate check*, page 111).
  - — Adjust if necessary.
- Remove and wash all deep rack rollers in warm water.
  - — Inspect for correct function, wear or damage.
  - — Rinse and re install.
- Check main drive shaft and drive chains.
- Carry out any other maintenance recommended by the manufacturer.
- Report all faults.

MONTHLY
- **Follow manufacturer's recommendations.**
- Inspect all racks and component parts, during cleaning.
- Clean filters.
- Drain all tanks, clean and re fill with fresh solutions. This may not be economically possible however.
  - — An alternative is to inspect the condition of the solutions and tanks and change as felt necessary.
- Carry out any other maintenance recommended by the manufacturer or felt necessary.
- Report all faults.

QUARTERLY
- **Follow manufacturer's recommendations.**
- Discard remaining chemicals in replenishment tanks.
- Wash out replenishment tanks and flush hoses.
- Re fill with fresh chemicals.
- Inspect electrical connections.
- Carry out any other maintenance recommended by the manufacturer.
- Report all faults.

YEARLY
- If you do not have an ongoing service contract with the supplier or their agent and carry out the routine maintenance yourself, it is advisable to have them service your unit yearly.

Record keeping
- It is essential to keep records of all quality control procedures, maintenance and repairs carried out.
- Record all parts purchased.
- Record costs.
- Continually review procedures, repairs, costs and quality of radiographs.

By maintaining and using your processor correctly there will be:

- less down time.
- less inconvenience.
- lower costs.
- greater through put.
- greater efficiency.
- greater job satisfaction.

It is up to you!

Electrical/mechanical systems check

Frequency of check
- Daily.
- Yearly by service engineer.

Items to be checked:
- Electrical connections and cables.
- Micro switches.
- Indicator lights.
- Audio signals.
- Heaters and thermostats.
- Motors.
- Roller drives.
- Pumps.
- Hoses.
- Readouts.

Temperature

Temperature of the developer is critical and is usually about 35–37 °C (see chemical and film manufacturer's recommendations).

Temperature of the fixer is less critical but should be similar to that of the developer.

Developer temperature should be tested regularly.

Temperature check

Frequency of check
- Daily.

Equipment required
- Thermometer—non-mercury (preferably digital).

Method
- Test at the same time each day, **when developer temperature has stabilized**.
- Place thermometer in developer.

- Read while in developer.
- Compare thermometer reading with the built-in readout temperature.

Evaluation
- Compare test temperature with:
  — Manufacturer's recommended temperature.
  — Built-in temperature.
- Record daily temperatures for comparison.
- Graphs may be plotted to show variations over time (see *Sensitometry*, page 112).
- look for variations of temperature:
  — Too high.
  — Too low.
  — Irregular.

Action
- Regular variations of more than one or two degrees should be investigated.
- Check for likely cause:
  — Heater.
  — Thermostat.
  — Tests done at different times of the day.
  — Developer temperature not reaching operating level after start up.
- If necessary, call the processor service technician.

Replenishment of developer and fixer

Similarly to manual replenishment, automatic replenishment maintains the level and activity of the processing chemistry. Pumps automatically replenish developer and fixer tanks each time a film enters the processor.

The replenishment rate is pre-set but can be varied within manufacturer's recommendation. It should be based on the area of film passing through the processor.

Replenishment rate test

Frequency of test
- Monthly.

Equipment required
- One graduated 100 ml cylinder.

Method
- Check that replenishment hoses do not have any bends that may restrict the flow.
- Switch on the processor.
- Remove lid of processor.
- Place activate micro switches, if present, to ensure that processor will function with the lid off.
- Place the end of the developer replenishment hose in the graduated cylinder.
- Feed a film into the processor.
- A pre-set amount of developer will automatically be pumped into the cylinder. Record this amount.
- Empty cylinder and repeat for fixer.

Evaluation
- Compare amount of developer and fixer in cylinder with manufacturer's recommendations.

Action
- If amounts do not compare with recommendations, adjust quantity regulator accordingly (see manufacturer's instructions).
- Re-check.
- File a report.

Cleaning

Adequate cleaning avoids the risk of film contamination, chemical build up on working parts and improves general health and safety.

DAILY

At start up
- Feed a 35 × 43 cm waste unprocessed film into the processor, to clean any deposits from rollers. It may be necessary to feed in a second film if you feel it is necessary. Do not use processed film as this will be too hard and may contain residual fixer.
- Do not proceed with the daily workload until you are satisfied the rollers are clean.

At shut down
- Remove and clean all crossover rollers or plates.
- Wipe off all chemical deposits on or around the processor. Take special care with film feed tray.
- Drain the wash tank.

WEEKLY
- Remove, clean and replace developer, fixer and wash roller racks.

MONTHLY
- Drain the system completely, including recirculation hoses and rinse with water.
- Charge the system with a dilute concentration of hypochlorite bleach (e.g. a 0.5% hypochlorite solution). **Take note of precautionary information on hypochlorite container.**

- Let the solution sit in the system for up to 30 minutes. Higher concentrations of hypochlorite or longer dwell times may harm some construction materials.
- Rinse the bleach from the system and dislodge easily removable biogrowth.
- Scrub the biogrowth off accessible surfaces. Use a clean, stiff brush or other tools recommended, for cleaning the surface.
- Flush the system thoroughly with water before returning it to normal use.

DO NOT allow concentrated sodium hypochlorite bleach to come in contact with photographic processing solutions. Dangerous fumes can be produced.

Summary of daily solution checks

Developer
- Check temperature at least once a day (ideally 3 times per day).
- Check developer level at time of temperature check.

Fixer
- Check fixer level at time of temperature check.

Water
- Check water level and flow at time of temperature check.

Sensitometry

Sensitometry is the study and measurement of the relationship between exposures, films, screens and processing.

Principle use
- Our interest lies more in its use in checking film processor performance, in particular that of automatic processing.
- By standardizing exposure, film and screen types, conditions under which films are exposed, handled and stored, leaves only one variable, that of film processing.
- Any variation in film image must therefore be due to film processing.
- A very elementary form of evaluating film processing performance has already been dealt with (see Module 4. Manual processing, page 78).
- Sensitometry is a more comprehensive and accurate form of monitoring processing performance.

When to do processor control sensitometry
- First thing every morning.
- After the processor has reached the correct operating temperature.
- After feeding clean up film through.
- Before processing any patient radiographs.
- After cleaning or servicing the processor.

Outline of procedure
- A standard step wedge image must be produced.
- This image consists of a range of clearly defined densities.
- Production of this image must be consistent. Methods for its production are described following.
- The film is processed in the film processor to be assessed.
- The resultant image densities are determined using a **densitometer**, recorded, graphed and the graphs evaluated.
- The first graph produced is known as a **characteristic curve**.
- To draw a characteristic curve, plot **test film densities** against **test film density step numbers**. Step 1 must be the lightest density.
- From this characteristic curve, several other graphs may be plotted, each giving additional information.
- This combined information will give a comprehensive picture of processor performance.

The test film

The image on the test film must be a standard series of clearly defined densities, ranging from barely visible to black.
- These densities are usually over a range of 21 steps.
- A smaller number of steps may be used.
- Number the steps from 1 (lightest step).

Producing the test film
There are several ways of preparing the test film.

METHOD 1
The best and most reliable method is to use a **sensitometer**, which produces a standard range of densities.

Using the sensitometer
- Use a dedicated box of film of the type commonly used in your department.

- Use the sensitometer in the darkroom where the processor is to be monitored.
- Select the light colour relevant to your films colour sensitivity (blue or green).
- Under safelight conditions insert a sheet of film into the sensitometer until it reaches the back-stop.
- Press cover down and hold until signal (audio or light) has stopped.
- Raise the cover and remove film.
- Process the film in the processor to be monitored.
- The film should always be placed in the same spot on the feed tray, with the image parallel to the rollers.

METHOD 2
Make an X-ray image of **aluminium step wedge** under standard conditions.

Making the step wedge image (see Fig 5–1)
- Place an 18 × 24 cm cassette, loaded with your standard film, face up on the X-ray table.
- Place the step wedge on the face of the cassette.
- Using a 100 cm FFD (SID), centre and collimate to the step wedge. The cassette can be divided with lead rubber in order to make more than one image on the same film.
- Set an exposure that will produce a full range of step wedge densities and make an exposure. You may need to experiment first in order to determine the correct exposure for your step wedge/film/screen combination.
- Process the film, under safelight conditions, in the processor to be monitored.
- The film should always be placed in the same spot on the feed tray, with the image parallel to the rollers.

Fig 5–1. Making a sensitometry test strip using an aluminium step wedge

- Standard conditions must be used each time a step wedge image is made.

METHOD 3
Producing a standard range of densities, using X-ray, **without** the step wedge.

Making the Image
- Place an 18 × 24 cm cassette, loaded with your standard film, face up on the X-ray table.
- Divide the face of the cassette into 11 strips.
- Cover all but the end strip with lead rubber.
- Set a 100 cm FFD (SID), centre and collimate to the uncovered area **(strip 1)**.
- Expose using a low exposure (enough to produce a barely visible density).
- Move the lead rubber so that **strips 1 and 2** are uncovered.
- Using the same exposure, expose both strips (strip 1 has now been exposed twice).
- Move the lead rubber so that **strips 1, 2 and 3** are uncovered and expose all three strips, using the same exposure (strip 1 has now been exposed three times and strip 2 twice).
- Repeat this process until all strips have been exposed.
- Process the film, under safelight conditions, in the processor to be monitored.
- Feed the film into the processor in the same way as that described in Method 2.
- The film may be cut down the middle lengthways, when removed from the cassette, in order to create a second strip. The unused strip should be placed in a light tight box for future use.

METHOD 4
- Purchase **pre-exposed sensitometry** film produced by a film manufacturer.

Terminology
To understand the process of sensitometry it is necessary to have an understanding of some basic terminology.

Sensitometer
A consistent light source, that produces a standard range of densities, when exposed on film.

Densitometer
A consistent light source, combined with a light measuring sensor, used for accurately measuring film density.

Film density
The degree of film blackening. You will see from the characteristic curve graph, and your own experience, that density increases as exposure increases.

Contrast
The difference between two or more densities on a film. The straight line portion and shape of the characteristic curve gives us information about contrast. A high contrast film curve will lie toward the left, whilst a lower contrast curve will lie more to the right (see exposure latitude, page 116).

Gradient
The contrast of a film at a given density. When a straight line is drawn tangent to the characteristic curve at a particular density, this line forms the slope which is the gradient of that density.

Average fradient
A line drawn between the **0.25 and 2.00 density levels** on the characteristic curve.

Toe gradient
A line drawn between the **0.25 and 1.00 density levels** on the characteristic curve.

Mid gradient
A line drawn between the **1.0 and 2.0 density levels** on the characteristic curve.

Upper gradient
A line drawn between the **2.0 and 3.0 density levels** on the characteristic curve.

Base plus fog
The density of a processed film, without the effects of light or radiation. The density at which the characteristic curve starts.

Exposure
Intensity of radiation × time (mAs).

Speed
Indicated by the location of the curve along the step (exposure) axis. A faster film curve will lie more toward the left, whilst a slower film will lie more toward the right. To calculate film speed use a density level of 1.0 (this is considered to be the average of the useful density range of 0.25 to 2.0).

Exposure latitude
The range of exposure factors, within which the resultant radiograph is considered to be acceptable. A film that is said to have a "wide latitude" has the ability to accept large changes in exposure, without excessive density changes.

Carrying out the sensitometric test

Frequency of test
- Daily.

Equipment required
- Sensitometer, to produce image by Method 1 described above, *or*
- Step wedge to produce image by Method 2 described above, *or*
- Lead rubber sheet and cassette to produce image by Method 3 described above.
- Unexposed film or manufacturer's pre exposed sensitometry film.
- Densitometer.
- Specialised graph paper supplied by a film manufacturer, *or*
- Simple small grid, graph paper, available from stationary shops, drawn up in the same way.
- Any scale graph paper may be used, but it is common to use a ratio, X axis to Y axis of 0.15: 1.0. **The important thing is that you do not vary the ratio once your quality control program is under way.**

Method
- Allow the processor to stabilise.
- Process the test film in the processor to be checked.
- Tests must be carried out at the same time each day under the same conditions.
- Developer temperature must be taken at the time of test.
- Using the **densitometer**, read the density of each step on the test film image.
- Record the densities against their step numbers.
- Plot a graph of **step numbers** on the **horizontal axis** against **densities** on the **vertical axis** and join up the plots (see Fig 5–2).

Using the densitometer

- Switch on the densitometer and allow unit to stabilise.
- Set densitometer readout to zero.
- Place the centre of the density to be read, directly over the aperture and under the reading arm.
- Lower the reading arm to the film, press the "read" switch and hold for a few seconds, until the readout stabilises.

MODULE 5. AUTOMATIC FILM PROCESSING

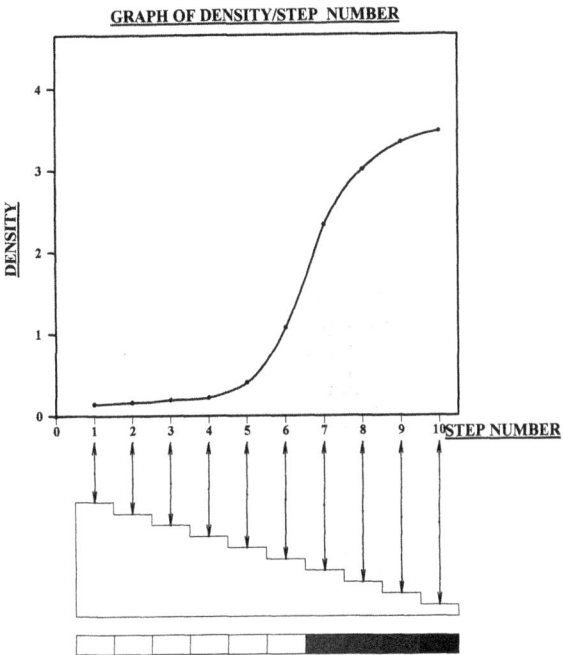

Fig 5–2. A typical characteristic graph showing relationship of density to step number

- Lift the arm.
- Record the reading against the step number.
- Repeat for all density steps on the image.

Plotting the characteristic curve
- The graph paper must have "**Density**" on the **Y (vertical) axis** and "**Step Number**" on the **X (horizontal) axis**.
- Plot density of step 1 (the lightest density) at the appropriate level on the Y axis and directly above Step 1 on the X axis.
- Repeat for all other step densities.
- Join up all plots to form a free flowing curve (see Fig 5–2).
- Draw in the **toe**, **mid** and **average** gradients as described above.

Evaluation
- Compare your characteristic curve with the control characteristic curve that was plotted when the chemistry was first mixed and the processor set up.
- If these curves vary markedly there is reason to believe that the processor is not functioning correctly.
- Calculate the "Speed", "Contrast", "Base plus Fog (D-Min)", "Temperature" and "D-Max" and enter on the appropriate graphs (see Fig 5–3 & Fig 5–4).

— Speed
  - Select the step number that has a density within a range of **1.0 to 1.3** on the step wedge image. This step number becomes the **speed step**.
  - Record the step number on the **speed chart** in the space provided (speed step No . . . .). Do this only on day 1, after the chemistry has been freshly mixed.
  - Plot the density reading on the zero line under Day 1. Do this only on day 1, after the chemistry has been freshly mixed.
  - Measure and plot the density of the Speed Steps daily (see Fig 5–3, Fig 5–4 and *Appendix B*, page 163).
  - Observe how much the speed plot varies from the zero line.
  - An acceptable variation is **plus or minus 0.15**.
  - **Variations above this may need corrective action.**
— Contrast
  - Select densities of the steps, two steps above and two steps below, the speed step.
  - Record these steps on the contrast chart, in the spaces provided (Step Below . . . . and Step Above . . . .).
  - Subtract the smaller from the greater of these two densities. This difference is the contrast of your standard reference.
  - Record this standard reference against the zero line on the **Contrast Chart**. Do this only on day 1, after the chemistry has been freshly mixed.
  - Repeat this procedure daily, plotting the contrast indicator under appropriate dates (see Fig 5–3, Fig 5–4 and *Appendix B*, page 163).
  - Observe how much the contrast indicator varies from the zero line.
  - An acceptable variation is **plus or minus 0.15**.
— Base plus fog (D-Min)
  - Using the densitometer, measure the density of the film that has received no exposure.
  - Record this reading against the zero line on the **Base Plus Fog Chart**. Do this only on day 1, after the chemistry has been freshly mixed.
  - Repeat this procedure daily, plotting the base plus fog density under appropriate dates (see Fig 5–3, Fig 5–4 and *Appendix B*, page 163).
  - Observe how much the base plus fog indicator varies from the zero line.
  - Ideally it should not go above **0.02. Action should be considered if it exceeds 0.023.**
— Temperature

Characteristic Curve

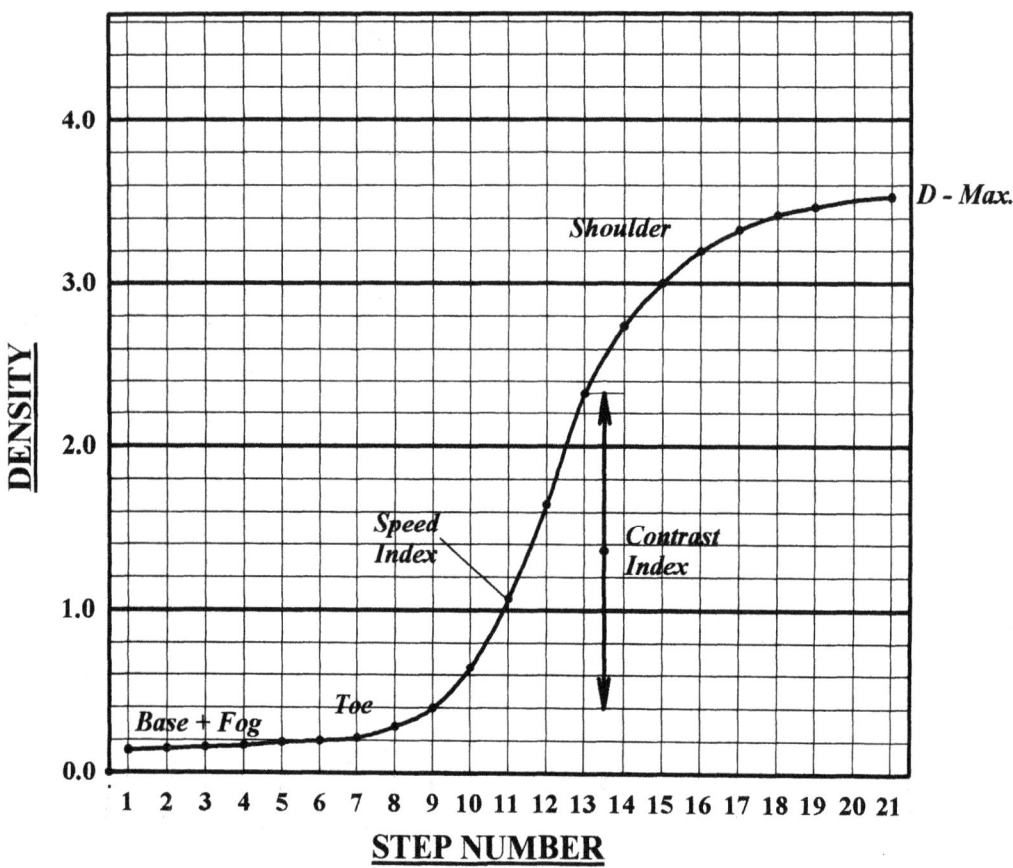

Fig 5–3. An example of a characteristic curve

- Take the temperature of the developer.
- Record against the zero line of the **Temperature Chart**. Do this only on day 1, after the chemistry has been freshly mixed.
- Repeat this procedure daily, plotting the Temperature under appropriate dates (see Fig 5–3 & Fig 5–4 and *Appendix B*, page 163).
- Observe how much the temperature varies.
- **Action should be considered if the temperature varies more than a few degrees.**
— D-Max
- Using the densitometer, measure the **maximum** density step on the step wedge image.
- Record this against the zero line on the D-Max Chart. Do this only on day 1, after the chemistry has been freshly mixed (see Fig 5–3, Fig 5–4 and *Appendix B*, page 163).
- Repeat this daily, recording the D-Max on the chart under appropriate dates.
- **Noticeable variations give advanced warning of chemistry problems.**

How to use the sensitometry graphs

All processors will show some variation in results from day to day. However, action should be taken if the variations are sudden or continue to increase or decrease over a period of time and pass beyond the acceptable limits.

MODULE 5. AUTOMATIC FILM PROCESSING

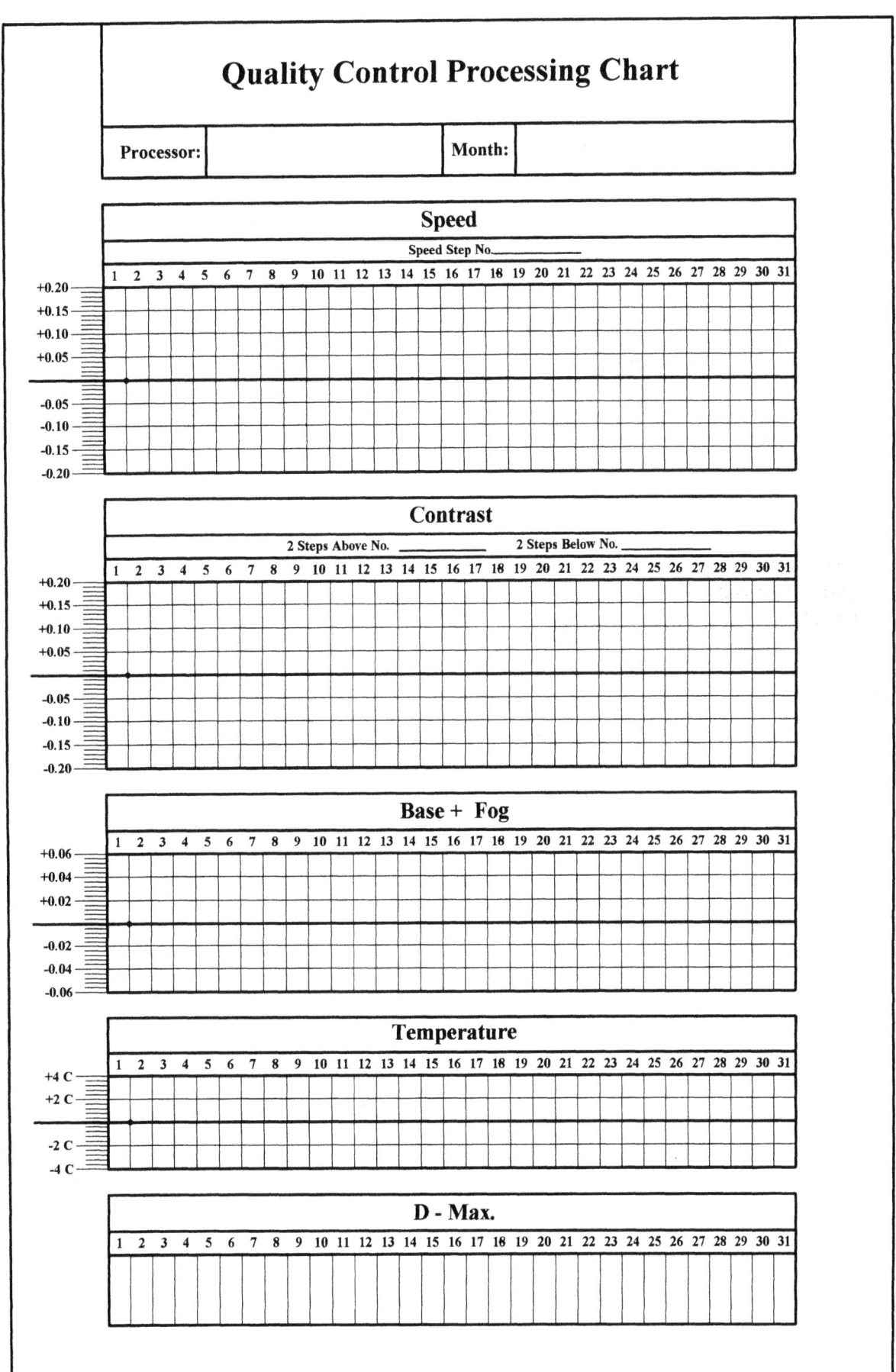

Fig 5–4. Sensitometry charts for recording Speed, Contrast, Base + Fog, Temperature and D-Max data

If the change is sudden, then all possible influencing factors should be checked, or a mistake may have been made. The test should be repeated.

In control:
When results remain within acceptable limits the processor is said to be "in control" and no action is necessary.

Out of control:
If one or more of the charts (speed, contrast or fog) is showing results that are outside the acceptable limits, especially if the change is sudden or continues to increase or decrease, then the processor is said to be "out of control" and immediate action should be taken.

Action to be taken if the processor is out of control
- Immediately stop using the processor.
- Inform other users
- Start problem solving process
- File a report

Chart changes	Possible cause	Action
• Speed and contrast increase • D min acceptable • Speed & contrast up • D min increases	*First stage of over development* 1 Developer temp. high 2 Excessive replenishment 3 Developer too concentrated 4 Check processing time	1 Adjust temp. 2 Adjust replenishment 3 Change developer 4 Developing time too long add starter
• Sudden increase in Speed, D min & D max after service	*Excessive over development* Starter omitted or insufficient Starter in developer	Add starter to developer
• Speed decreases • Loss of image contrast • D min normal • Image density low over whole image Speed & contrast low • D min also low	*Under development* 1 Developer temp. low 2 Developer exhausted 3 Insufficient replenishment 4 Replenisher used up 5 Developer too dilute 6 Developing time too short	1 Check & adjust temp. 2 Check & adjust replenishment 3 Check & refill replenisher tank 4 Replace replenisher 5 Check processing time
• Sudden decrease in speed and D max after service • Small decrease in D min	Excessive amount of starter in developer	Replace developer, add correct amount of starter
• Increase in speed • Shoulder decreases • Loss of contrast • Increase in fog	1 Aerial oxidation 2 Contaminated developer	Replace developer after washing out tank Add correct amount of starter

Fig 5–5. Example of the possible interpretation of sensitometric charts and recommended action.

Notes

TASK 24
Sensitometry test

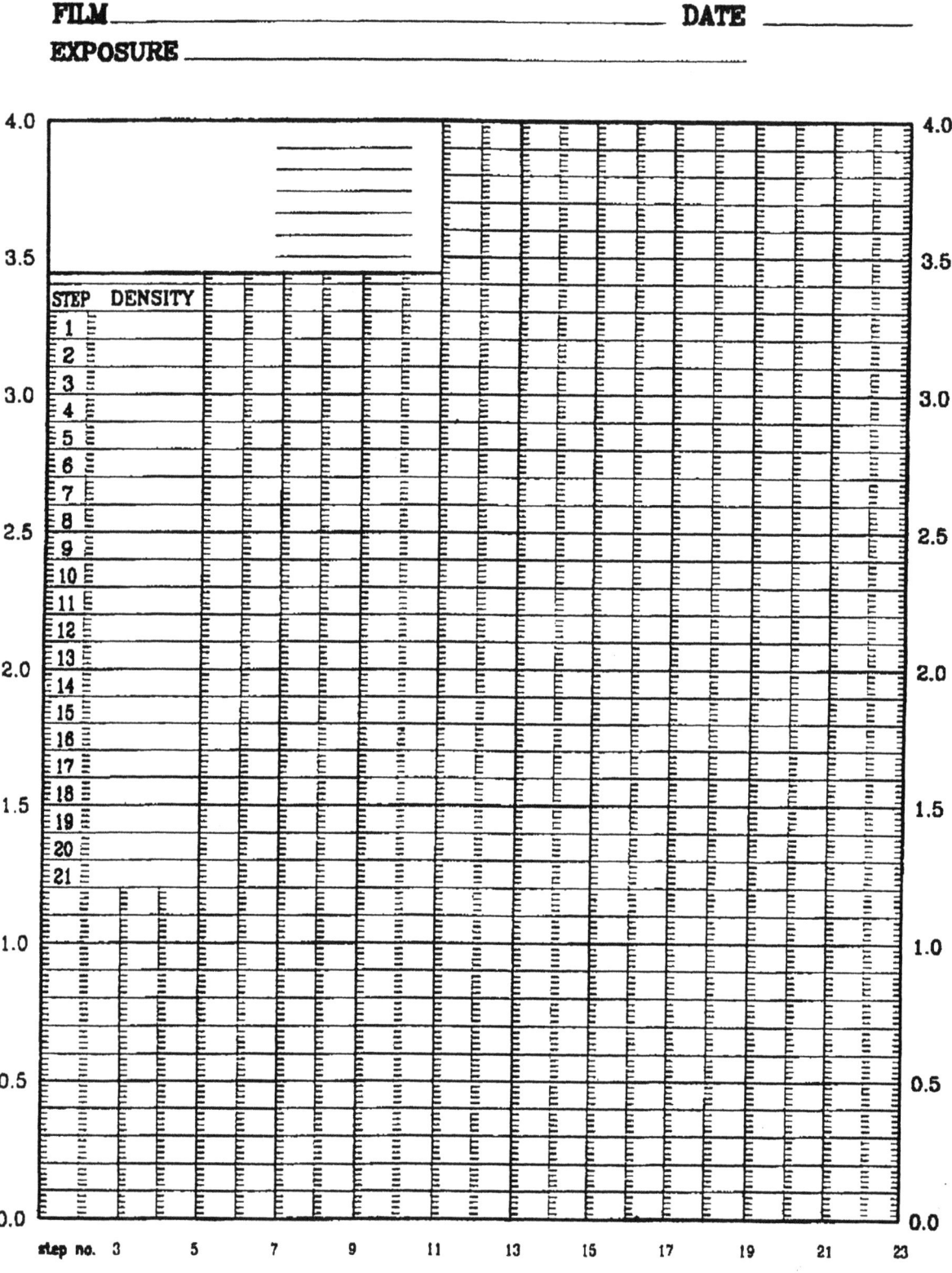

MODULE 5. AUTOMATIC FILM PROCESSING

You are required to carry out the daily monitoring sensitometry check of your processor.

    a) Select an automatic processor.
    b) Carry out the routine sensitometry test.
    c) Plot the Characteristic Curve on the graph sheet below.
    d) Answer the following questions in the spaces provided.

1. How does this graph compare with the Characteristic Curve shown in the notes?

   _____
   _____
   _____
   _____
   _____

2. Give the following figures: SPEED _____
                                       CONTRAST _____
                                       BASE + FOG _____
                                       D-MAX _____

Include your step wedge film with these answers.

Tutor's comments:

Satisfactory/Unsatisfactory

Signed _____   Date _____
                Tutor

QUALITY ASSURANCE WORKBOOK

TASK 25
Automatic processor start up routine

You have been asked to start up the automatic processor at the beginning of a working day.

a) Select an automatic processor.
b) Carry out the automatic processor start up routine using the following check list.
c) Answer the following questions in the spaces provided.

Check	Satisfactory/Unsatisfactory	Comment
Developer level		
Developer replenisher level		
Developer replenishment rate		
Developer condition		
Developer temperature		
Fixer level		
Fixer replenisher level		
Fixer replenishment rate		
Fixer condition		
Water flow		
Condition of rollers		
Condition of drive cogs		
Drive motor		
Dryer		

MODULE 5. AUTOMATIC FILM PROCESSING

Feed warning signal
_____
_____

Passage of clean up film
_____

Abnormal noises
_____

Cleanliness
_____

1. What was the developer replenishment rate? _____

2. What was the fixer replenishment rate? _____

3. What was the developer temperature? _____

4. What are your recommendations?
_____
_____
_____
_____

_____

Tutor's comments:

                                                    Satisfactory/Unsatisfactory
_____

Signed _____   Date _____
            Tutor

MODULE 6
Radiographic exposures

Aim

To provide sufficient knowledge of exposure selection and manipulation, to enable accurate selection of exposures, in order to produce maximum information on the resultant radiographs.

Objectives

On completion of this module the student will

- Understand the effect each exposure factor has on the resultant radiograph.
- Be able to use this knowledge effectively.
- Be able to select the correct exposure factors required.
- Understand and use the **step system** of exposure calculation.
- Be able to manipulate exposure factors using the **step system**.
- Be able to establish and maintain a reliable exposure chart.
- Be able to modify an existing exposure chart.

Strictly speaking the term, "exposure" refers to the quantity of radiation to which the patient is exposed.

In practical radiography we tend to talk of "exposure" as the collective exposure factors, kV, mA and time, which together will produce the radiation exposure that will give the required penetration, density and contrast on the radiograph.

There are several other factors, which influence film quality, but here we will only consider kV, mA, time and FFD (SID).

kV

- Controls largely the **penetrating power** (better described as the quality of the beam), and to a lesser degree, the intensity of radiation and therefore film density and patient dose.
- Affects **contrast**. The higher the kV the lower the contrast. The lower the kV the higher the contrast.

mA

- Controls the **intensity** of radiation and therefore film density and patient dose. The higher the mA, the higher the film density and patient dose.

time

- Controls the **length of time** of radiation flow, and therefore film density and patient dose. The longer the exposure time the higher the film density and patient dose.

mAs

- The multiple of **mA and time** (in seconds).
- Many modern X-ray units are designed to use mAs rather than mA and time separately.

FFD (SID)

- The greater the distance an X-ray beam travels the less effective it will be, the less distance it travels the more effective it will be.
- Exposure compensation is therefore necessary with changes in FFD (SID).

For the radiographer it is important to set the correct exposure factors required to produce a high quality radiograph, which will give the maximum amount of information yet minimise the dose to the patient.

It is important to have systems in place that will help the radiographer to select and manipulate exposure factors effectively.

Exposure chart

Each X-ray unit should have a list of commonly used exposures for easy reference (see *Appendix B*, page 164). This list will, of course, be limited, but will at least provide a guide.

Methods of recording exposures
- Written chart.
- Notebook.
- Computerised "chart" within the X-ray unit.

MODULE 6. RADIOGRAPHIC EXPOSURES

Here we will deal with:
- Establishing and modifying an exposure chart.
- Manipulating exposure factors using the step system.

Establishing an exposure chart
Method
- Draw up a blank exposure chart (see Fig 6–1 and *Appendix B*, page 168).
- Fill in all the anatomical areas to be covered, listing the views for each.
- Divide these into groups using the same set of conditions. e.g. all extremities using detail screens, no grid, 100 cm FFD (SID).
- Produce a well exposed set of radiographs of an extremity, e.g. hand. (this can be done in the course of a routine examination.)
- Record the exposures used against HAND on the chart.
- Measure the thickness of the patient's hand in all projections used. e.g. PA, Oblique, lateral (see *Module 2. Accessory equipment*, page 41 and *Appendix A*, page 135).
- Measure thickness at the level of entry of the central ray, for each position of the hand.
- Measure all other areas/positions in the same group (these can be obtained from a colleague or friend of similar size rather than inconvenience the patient).
- Calculate the exposures for all other areas/positions in the group, based on the following chart (patient thickness related to exposure change), using the hand as your base level:

Patient thickness related to exposure change
— 1.5 cm increase in thickness requires a 25% increase in exposure (+1 step)
— 5.0 cm increase in thickness requires a 100% increase in exposure (+3 steps)

Exposure chart

Room _____

Area	kV	mAs mA	Time	FFD (SID) cm	Grid	Screens	Remarks
Hand							
PA				100	—	Detail	
Oblique				100	—	Detail	
Lateral				100	—	Detail	
Wrist							
PA				100	—	Detail	
Oblique				100	—	Detail	
Lateral				100	—	Detail	
Forearm							
AP				100	—	Detail	
Lateral				100	—	Detail	
Elbow							
AP				100	—	Detail	
Oblique				100	—	Detail	
Lateral				100	—	Detail	
Foot							
DP				100	—	Detail	
Oblique				100	—	Detail	
Lateral				100	—	Detail	
Ankle							
AP				100	—	Detail	
Oblique				100	—	Detail	
Lateral				100	—	Detail	

Fig 6–1. An example of a partially completed exposure chart

QUALITY ASSURANCE WORKBOOK

- 5.0 cm decrease in thickness requires a 50% decrease in exposure (–3 steps)
- 1.5 cm decrease in thickness requires a 23% decrease in exposure (–1 step)

Note: The reference to "+ or – steps" is explained below *The step system*.

Example
- PA HAND (**2 cm thick**) Exposure **50 kV 6 mAs**.
- Calculate an exposure for a lateral WRIST (**7 cm thick**).
- Thickness difference between PA HAND and Lateral WRIST is **5 cm**.
- For an increase of **5 cm** the exposure is increased by 100% (see "Patient thickness related to exposure change", figures above).
- Therefore the lateral WRIST exposure will be **50 kV 12 mAs**.
- The exposures for all other areas of the body can be calculated in this way.

The step system

The step system is a simple, standardised form of exposure factor manipulation, designed to remove some of the guess work and make exposure setting more accurate.

It works on a series of step charts allowing the use of standard step changes in exposure.

It is also related to patient thickness (see *Appendix A*, pages 135 & 136) and medical condition (see Fig 6–5).

Note: Step changes can be applied to any one of the charts or split between the charts.

- The step system provides step charts for kV, mA, mAs, time and FFD (SID), (see Fig 6–2, Fig 6–3, Fig 6–4 & Fig 6–5).
- To use the step system to make exposure changes, it is necessary to talk in terms of **steps**.
- **Each step** on any of the charts will alter the exposure by approximately **25%**.
- For a **noticeable change** in film density to be seen, the exposure must be altered by **at least 25%**.
- If a film is considered to be **too light or too dark** it will be necessary to change one of the exposure factors by **at least three steps** up or down.

kV
40
41
44
46
48
50
52
55
57
60
63
66
70
73
77
81
85
90
96
102
109
117

Fig 6–2. kV step chart

mAs	mA	seconds
		.010
		.012
		.016
2	20	.020
2.5	25	.025
3.2	32	.032
4	40	.040
5	50	.050
6.4	64	.064
8	80	.080
10	100	.100
12.5	125	.125
16	160	.160
20	200	.200
25	250	.250
32	320	.320
40	400	.400
50	500	.500
64	640	.640
80	800	.800
100	1000	1.00
125		1.25
160		1.60
200		2.00
250		2.50
320		3.25
400		4.00
		5.00

Fig 6–3. mA, mAs and time step charts.

FFD (SID) cm
200
178
158
140
125
112
100
89
80
71

Fig 6–4. FFD (SID) step chart

- Any **step change** required to modify an exposure can be applied to any of the Step Charts, although other factors will influence this decision (contrast, penetration, patient dose, movement, limitations of the X-ray unit).
- Step changes need not be limited to one chart only. The necessary step changes can be split up between charts if necessary, providing that the overall number of required step changes is made.

Example
- Original Exposure = 60 kV 20 mAs
- New kV = 70 kV (an *increase* of 10 kV)
- An *increase* of 10 kV = *three steps* on the kV step chart.
- To retain the same density film, the *mAs* must be *reduced* by the same number of steps (*three steps*) on the *mAs chart*.
- New Exposure = 70 kV 10 mAs

Step charts

All step charts reproduced from "Course Notes in Basic Radiography", Radiation Protection Branch, South Australian Health Commission.

Exposure modification for changes in conditions

Pathology	Step changes suggested
Ascites	+2
Asthmatic	-1 or 2
Bowel Obstruction	-1 or 2
Emaciated	0
Emphysema	-1 or 2
"Flabby" Fat	-1
Muscular Physique	+1
Osteoporosis	-1 or 2
Paget's Disease	+1 or 2
Pleural Effusion	+1 or 2
Pneumonia	+1 or 2
Pulmonary Oedema	+1 or 2
Tuberculosis	+1
Plaster of paris	
Half	+1 or 2
Dry	+3
Wet	+4 or 5
Fibreglass	0 or +1
Grid	
Up to Ratio 8.1, 40 lines/cm at 80 kV	+5
Screens	
Detail to Fast	-3
Fast to Detail	+3 (ideally use mAs steps)

Note: These step changes are a guide only. You may need to modify this chart to suit your own conditions.

Fig 6–5. Suggested step changes for specific changes in conditions

Notes

TASK 26
Radiographic exposures

You have been asked to complete the exposure chart shown below

a) Calculate all other exposures on the chart, based on the information given against DP Foot and AP Hip.
b) Fill in all other relevant information.

Exposure chart

Area	kV	mAs	Thickness of Part cm	FFD (SID) cm	Grid	Screens
Foot DP Oblique Lateral	50	10	8	100	————	Detail
Ankle AP Oblique Lateral						
Tibia & fibula AP Lateral						
Knee AP Lateral						
Femur (Lower 2/3) AP Lateral						
Hip AP Lateral	70	50	18	100	Bucky	Fast
Pelvis AP						

Tutor's comments:

Satisfactory/Unsatisfactory

Signed _____ Date _____
          Tutor

TASK 27
Modify an exposure chart

Your department has just been supplied with new intensifying screens and your exposure charts are no longer accurate. Someone has worked out some exposures already and has asked you to modify the remainder.

a) Modify the following exposure chart.
b) Start by replacing the relevant exposures with the following exposures.
   HAND, PA   50 kV 8 mAs
   SHOULDER, AP   65 kV 16 mAs
   CHEST, PA   100 kV 5 mAs
c) Now modify all other exposures on the chart, using the step charts.
d) Enter the new exposures alongside the old ones.

Exposure chart

Area	kV	mAs	FFD cm (SID)	Grid	Screen
Hand					
PA	50	16	100	No	Detail
Oblique	50	20	100	No	Detail
Lateral	50	32	100	No	Detail
Wrist					
PA	50	16	100	No	Detail
Oblique	50	20	100	No	Detail
Lateral	50	25	100	No	Detail
Forearm					
AP	55	20	100	No	Detail
Lateral	55	25	100	No	Detail
Elbow					
AP	55	25	100	No	Detail
Oblique	55	25	100	No	Detail
Lateral	60	20	100	No	Detail
Humerus					
AP	60	32	100	Bucky	Fast
Lateral	60	32	100	Bucky	Fast
Shoulder					
AP	65	32	100	Bucky	Fast
SI	65	32	100	Bucky	Fast
Chest					
PA	90	5	180	Bucky	Fast
Oblique	90	5	180	Bucky	Fast
Lateral	100	5	180	Bucky	Fast

MODULE 6. RADIOGRAPHIC EXPOSURES

Tutor's comments:

Satisfactory/Unsatisfactory

Signed _____   Date _____
          Tutor

TASK 28
Creating a new exposure chart

Your department has just been equipped with a new X-ray unit and you have been asked to produce an exposure chart for it.

- a) Select an X-ray room.
- b) Using the knowledge you have gained to date, establish a completely new exposure chart, for the X-ray unit in that room.
- c) Draw up a chart and fill in all relevant information on your chart. e.g. patient positions, exposure factors, grid, screens, FFD (SID), patient size.
- d) Submit the chart to your tutor for assessment.

Tutor's comments:

Satisfactory/Unsatisfactory

Signed   Date
            Tutor

APPENDIX A
Making simple test tools

Water phantom

Purpose
- To simulate a patient.
- To provide an overall even image.

Equipment required
- 20 litre plastic flagon (other plastic containers of a suitable size can be used).
- Ruler.
- Thick tipped waterproof marking pen.

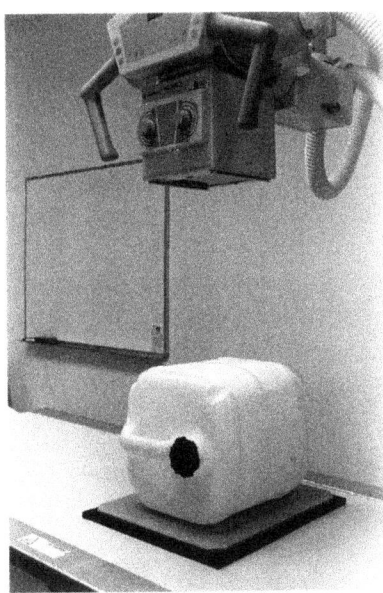

Fig A–1. Plastic flagon being used as an X-ray phantom

Method
- Lie the container on its side
- Using the pen and the ruler, mark the side of the container with various depths above the lower side of the container (10 cm, 20 cm and 30 cm).
- The container can be filled with varying depths of water, depending on need.

Note: Other items that can be used:
- Aluminium or wax block.
- Several books or phone directories.
- A pile of particle board (chip board) sheets.

Aluminium step wedge

Purpose
- To produce a standard range of densities on film.
- Sensitometry.
- Film comparison.
- Generator output consistency.
- mAs consistency.

METHOD 1

Equipment required
- Aluminium block, 105 mm long × 42 mm high.
- Hacksaw.
- File.
- Pencil.
- Rule.

Method
- Draft a template and stick it on the side of the aluminium wedge or draft the measurements directly on to the aluminium
- Cut or file the sloping surface of the wedge into **21 steps, each 5 mm long and 2 mm high**.
- Check that the height of each step is accurate.
- Smooth off rough edges.

Fig A–2. Aluminium step wedge

METHOD 2

Equipment required
- An aluminium strip **1.2 meters long 10 mm wide and 2 mm thick**.
- Adhesive suitable for gluing aluminium.
- Hacksaw.
- File.
- Pencil.
- Rule.

Method
- Cut the aluminium strip into **21 shorter strips**.
- The **first** strip must be **105 mm long**.
- The remaining 20 strips must be **cut progressively 5 mm shorter** than the preceding strip, making the last strip 5 mm long.
- There may be a little wastage.
- Place the strips one on top of the other, starting with the longest and getting progressively smaller until a series of even steps have been built up.
- Glue the strips together.
- Place the vertical end of the wedge against a firm upright surface and tap each strip back against it.
- Leave to harden.
- Smooth off the rough edges.

Note:
- The dimensions of the step wedge are not critical providing, the steps are standard.
- A simple step wedge can be made with only 11 steps.
- The steps need not necessarily be glued, but are less likely to get lost if they are.

Film/screen contact test tool

Purpose
- To test for poor film/screen contact within a cassette.

Equipment required
- A sheet of **fine gauge wire mesh** or sheet of thin **perforated zinc** (*do not use aluminium*).
- The size of the sheet should be slightly larger than a **35 × 43 cm cassette**.
- An old picture frame with an internal measurement slightly larger than a 35 × 43 cm cassette, or timber to make a frame.
- The frame and wire mesh/perforated zinc must be compatible in size.
- Box of large headed tacks, drawing pins or staples.

Method
- Make up the frame if necessary.
- Tack the wire mesh/perforated sheet to the frame so that it is perfectly flat.

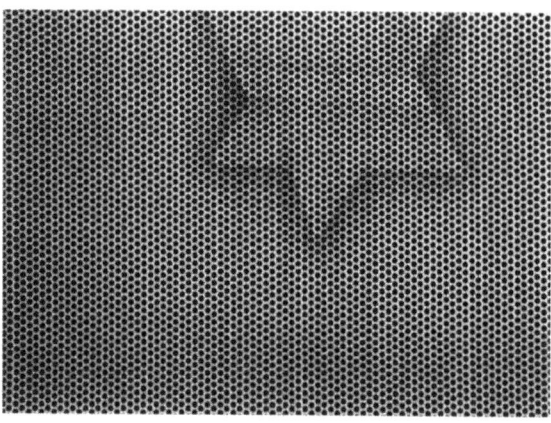

Fig A–3. Perforated zinc sheet—used as film/screen contact test tool

Note:
The wire mesh/perforated sheet can be used without the frame, provided it is kept perfectly flat. Care must be taken not to bend or damage it. Plastic sheets can be mounted either side to protect it.

A small hole, approximately 2 cm square, sited about 10 cm from one edge of the sheet helps to gauge the correct exposure and film density (densitometer readings would be taken here).

Spinning top timer test tool

Purpose
- Check accuracy of X-ray unit timer.

Equipment required
- Brass or mild steel disc, **100 mm in diameter, 2–3 mm thick**.
- Steel disc **60 mm in diameter, 10 mm thick**.
- Small steel block **20 mm × 20 mm × 20 mm**.
- Bolt **3 mm thick**, **40 mm long**, with 2 nuts and a countersink head.
- Drill and 3 metal drill bits, 3 mm, 5 mm and a countersink bit.
- Spanner.
- File.

APPENDIX A. MAKING SIMPLE TEST TOOLS

- Pencil.
- Rule.
- Welding outfit.

Method 1
- Weld the small steel block to the centre of the large diameter disc.
- Drill a **5 mm** hole through the centre of the larger disc and **15 mm** up into the small block. (do not drill right through the small block).
- File all sharp edges off the small block.
- Drill a **3 mm** hole in the large disc **10 mm** from its outer edge.
- Drill a **3 mm** hole right through the centre of the smaller disc and countersink one end of the hole.
- Insert the bolt into the hole in the smaller disc, so that the head is countersunk below the surface of the disc and fix firmly in place with both nuts.
- File the tip of the bolt to a point.
- Place the smaller disc on a firm surface with the bolt up.
- Hold the larger disc by the small welded block and fit the sharpened bolt into the central hole in the larger disc.
- By holding the small welded block the larger disc can be spun.

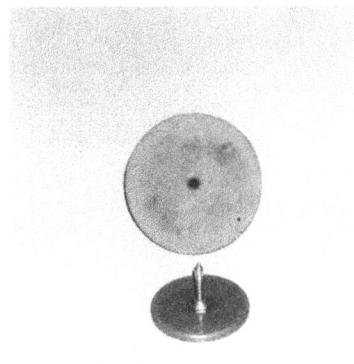

Fig A–5. A simple brass spinning top, (a) assembled ready for use, (b) with top lifted to show base

Measuring callipers

Purpose
- Measuring patient thickness in order to calculate exposure (see *Module 6. Radiographic exposures*, page 124).

Equipment required
- 2 strips of aluminium **600 mm long**, **20 mm wide**, **2 mm thick** (mild steel may be used, but it is more difficult to work with).
- Drill and 2 mm metal drill bit (or felt tip pen).
- Hacksaw.
- File.
- Pencil.
- Ruler.

Fig A–4. A commercially made spinning top

Method 2
- A wooden block or section of broom handle can be used instead of the small steel block, previously described, and fixed in place with two screws, after first drilling the disc to accommodate the screws.
- A wooden base 150 mm square can be used instead of the smaller of the metal discs.

Method
- Form a flattened loop at one end of one of the strips, forming a T shape.

- This flattened loop should fit snugly over the other strip and slide up and down it.
- Bend the second strip to form a **90° angle**.
- File all rough edges smooth.
- Slide the flattened loop over one end of the other strip of metal.
- Adjust so that the one strip slides easily up and down the other.
- Check that the two opposing arms are parallel to one another.
- Measure the distances in centimetres between the bottom of the adjustable arm and the opposing fixed arm.
- Mark these distances on the arm that the moveable arm slides on.
- A hacksaw blade can be used to mark the metal, or drill small holes.
- The numbers can be written against the marks with felt tip pen.

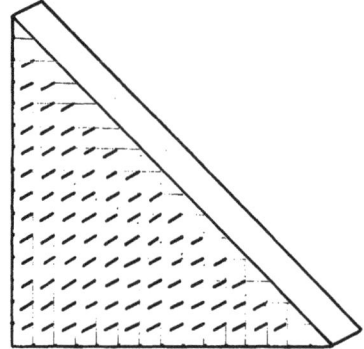

Fig A–7. Diagram of a the tomography test tool described in Method 1

Method
- Accurately measure from the base of the block, up the vertical side, marking the block every 1 cm.
- Repeat along the base.
- Draw in the intersecting lines.
- Partially hammer in nails, to the same depth, at the intersection of every line.
- It is important that the height of each nail above the base is accurate.
- Mark each horizontal line with its height above the base.
- On the resultant film the nails/pins that are clearly defined are those at the level of the tomographic cut.

Note: A polystyrene block and sewing pins can be used instead of the wood and nails.

METHOD 2

Equipment required
- Strip of fine metal gauze (the dimensions of the strip of gauze should match those of the sloping surface of the wedge).
- Set of small lead numbers 1 to 9.
- Small tube of adhesive.
- Rule.
- Pen or pencil.

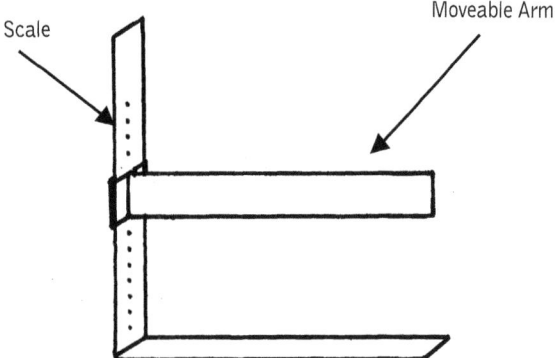

Fig A–6. Diagram of simple measuring callipers

Tomography test tool

Purpose
- To check the accuracy of tomography cut levels

METHOD 1

Equipment required
- A wooden wedge 15 cm base × 15 cm vertical × 21 cm slope and approximately 3 cm wide.
- Ruler.
- Pen or pencil.
- 91 small nails 2–3 cm long.
- Hammer.

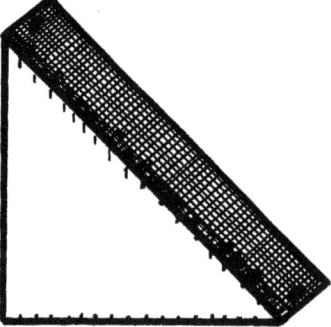

Fig A–8. Diagram of the tomography test tool described in Method 2

Method

- Measure and mark every 1 cm along the base, as in method 1.
- Draw in the vertical lines and measure the height of each one.
- Write the height against each line.
- Place the metal gauze on the sloping surface and fix with adhesive.
- Glue the lead numbers on the surface of the gauze, keeping them all toward the outer edge on the same side.

Note:
1. A polystyrene block can be used instead of wood.
2. If difficulty is experienced in producing a clear enough image of the mesh, try placing a strip of thin lead (the thickness of that used in the back of cassettes), or copper, under the test tool.

X-ray beam/grid alignment test tool

Purpose
- To check the alignment of the X-ray beam to the grid lines.

Equipment required
- Sheet of lead or lead rubber, 2 to 3 mm lead equivalent.
- Drill and two drill bits, one 2 mm and one 10 mm.
- Knife or hacksaw.

Method
- Cut the lead into one strip measuring 230 × 50 mm and two strips measuring 90 × 50 mm each.
- Drill five 10 mm holes in the large lead strip.
- Drill two 2 mm holes in the large lead strip, one either side of the central 10 mm hole.
- Drill a group of three 2 mm holes at one end of the large lead strip.
- See diagram below for measurements.

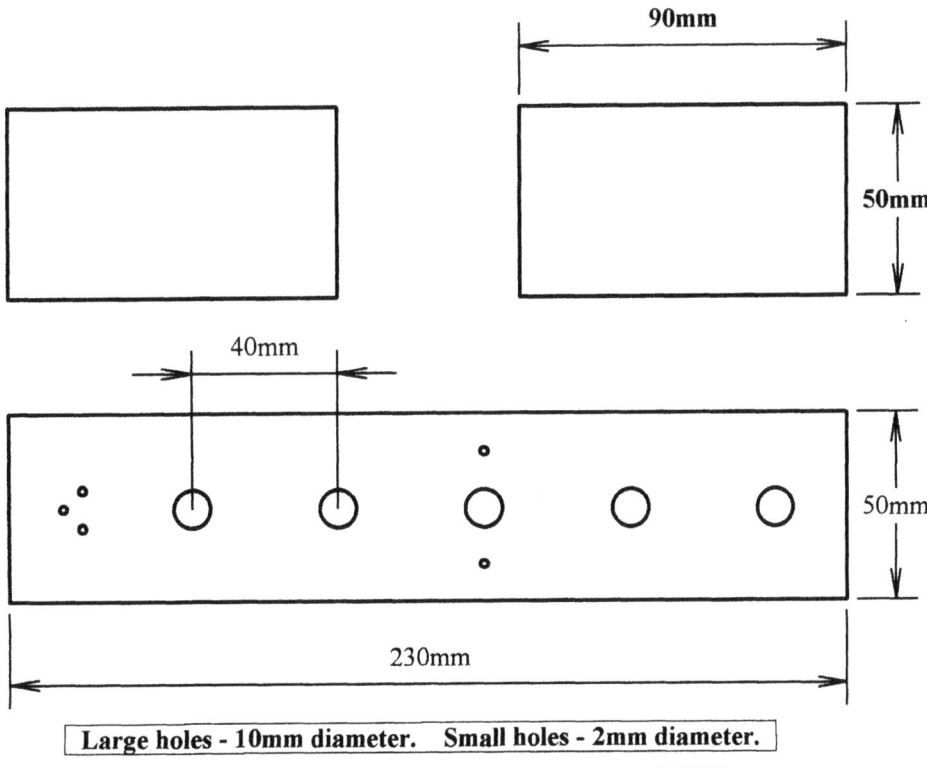

Large holes - 10mm diameter.   Small holes - 2mm diameter.

Material: 2mm Lead sheet or Lead rubber sheet.

X - ray Beam/Grid Alignment Test Tool.

Fig A–9. Diagram of an X-ray beam/grid alignment test tool with dimensions

APPENDIX B
Graphs, check sheets & record sheets

Reject film analysis
Daily totals

Period: From _____ To _____

Cause	Daily totals							Weekly totals
Dates								
Projection								
Movement								
Under exp.								
Over exp.								
Static								
Fog (darkroom)								
Fog (cassette)								
Equipment								
Processing								
Other								
Totals								

Total film rejected: _____
Total film used: _____
Rejected film as a percentage of film used: _____
Cost of film rejected this week: _____

Comments/action

Reject film analysis
Daily totals by film size

Date: _____

	Film size cm	No. of films	Amount of film (sq. m)
1	35 x 43		
2	35 x 35		
3	30 x 40		
4	24 x 30		
5	18 x 24		
6	18 x 43		
7	Dent / Occ		
8	Other		
	Totals		

Total number of films: _____
Total square meters of film : _____
Cost of rejected film today: _____
_____

Comments/action

Reject film analysis
Daily totals by location

Date: _____

Cause	Location				
	Room 1	Room 2	Room 3	Room 4	Total
Projection					
Movement					
Under exposed					
Over exposed					
Static					
Fog (darkroom)					
Fog (cassette)					
Equipment					
Processing					
Other					
Total					

Total films rejected today: _____

Comments/action

Remedial action record
Following reject film analysis

Date reject film analysis completed _____

Date started	Remedial action	Date finished	Outcome

Collimator check records

Room _____ FFD (SID) _____ Light field size _____
kV _____ mA _____ Sec _____ mAs _____

Date	Light/X-ray coincidence test A B C D	Visual inspection	Electrical inspection	Shutter efficiency	Remarks

- The letters A,B,C,D, under "Light/X-ray coincidence test" above, refer to individual collimator shutters, "A" being the shutter on the control knob side, then rotating clockwise.
- Record the difference in mm between the light and X-ray images, indicate if the light field is **outside** or **inside** the X-ray field.

Record of all cassettes

No.	Size cm	Type of screens	Date fitted	Date cassette put into service

Cassettes and screens
Maintenance and testing

Record dates of maintenance/tests

Cassette No.	Screen comparison test	Film/screen contact test	Screen cleaning	Cassette maint.	Cassette light tight test	Comment

APPENDIX B. GRAPHS, CHECK SHEETS & RECORD SHEETS

Cassettes and screens
Routine maintenance and cleaning record

Date	Cassette numbers	Comments

Lead rubber aprons and gloves
Testing record

Dates inspected													
Fasteners													
Cover													
Stitching													
X-ray test													
Hangers													
Are hangers used?													
Cleanliness													

- Enter the date checks carried out
- Place a tick in the box if satisfactory
- Place a cross in the box if unsatisfactory and enter comment below

Comments: (Enter date, comment and signature)

Date							Comment							Signature

_____
_____
_____
_____
_____
_____
_____
_____
_____
_____
_____
_____
_____
_____

Viewing box
Maintenance and inspection

Dates inspected													
Position													
Safely fixed													
Light — even — brightness													
Electrical — plug — wiring — switch — tubes — starters													
Window													
Film clip													
Cleanliness — inside — outside													
Other													

Comments: (Enter date and signature)

Date   Comment   Signature

_____
_____
_____
_____
_____
_____

Fault report
To be completed by person reporting fault

Name	Time
	Date
Room	Priority (circle a number)
	1 2 3 4 5 6 7 8 9 10
	urgent not urgent

Description of problem (Please be specific)

Signed _____

Action

Signed _____

Outcome

Signed _____

Equipment record

Room No. _____

X-ray generator

Manufacturer _____   Supplier _____

Model _____   Serial No. _____

Date installed _____   Date accepted _____

Warranty expiry date _____

Sing phase _____   Three phase _____

Maximum kVp _____   Maximum mA _____

X-ray tube

Manufacturer _____   Supplier _____

Model _____   Serial No. _____

Focus sizes
    Broad _____ Fine _____   Warranty expiry date _____

Date installed _____   Date accepted _____

Tube suspension

Manufacturer _____   Supplier _____

Model _____   Serial No. _____

Mounting—Floor/Ceiling _____   Locks—Manual/Magnetic

Date installed _____   Date accepted _____

Warranty expiry date _____

Collimator

Manufacturer _____ Supplier _____

Model No. _____ Serial No. _____

Light switch—Clockwork/Electronic Warranty expiry date _____

Date installed _____ Date accepted _____

Table

Manufacturer _____ Supplier _____

Model No. _____ Serial No. _____

Movement—Tilt/Fixed Potter–Bucky—YES/NO

Potter-bucky grid
   Ratio _____ Line _____ Focal range _____
   Movement—YES/NO

Upright potter-bucky

Manufacturer _____ Supplier _____

Model No. _____ Serial No. _____

Grid Ratio _____ Grid line _____

Focal range _____ Grid movement—YES/NO

Date installed _____ Date accepted _____

Warranty expiry date _____

Control panel

Manufacturer _____ Supplier _____

Model No. _____ Serial No. _____

Date installed _____ Date accepted _____

Warranty expiry Date _____

Stationary grids

Manufacturer _____   _____   _____

Supplier   _____   _____   _____

Model   _____   _____   _____

Serial No.   _____   _____   _____

Grid ratio   _____   _____   _____

Grid line   _____   _____   _____

Focal range   _____   _____   _____

Date supplied _____   _____   _____

Notes

_____
_____
_____
_____
_____
_____
_____
_____
_____
_____
_____
_____
_____
_____
_____
_____
_____
_____
_____
_____
_____
_____
_____

Equipment maintenance and repair log

Date	Action	Repair hours	Service hours	QC hours

X-ray unit
Record of visual/manual quality control checks

Room No. _____

Pass = ✓

Fail = X

Does not apply = NA

	Date									
Tube and tube suspension	FFD Scales									
	Angulation indicator									
	Locks (all)									
	Perpendicularity									
	Collimator									
	Tracks									
	High tension cable/other cables									
	Cleanliness									
Table and upright bucky	Bucky mobility									
	Bucky lock									
	Cassette lock									
	Bucky grid movement									
	Cables									
	Table									
	Cleanliness									
Control panel	Hand switch and cable									
	Panel switches, lights and meters									
	Technique charts									
	Overload protection									
	Cleanliness									

Fluoro. system	Locks (all)										
	Power assist										
	Motion smoothness										
	Switches/lights/meters										
	Compression device										
	Fluoro. monitor										
	Fluoro. grid										
	Fluoro. timer										
	Fluoro. shutters visible										
	Cleanliness										
Other	Gonad shield, aprons and gloves										
	Measuring callipers										
	Stationary Grids										
	Patient Positioning Aids										

Constancy of radiation output at different mA settings
Test record

Room _____ X-ray Unit _____
kV _____ Focal Spot. Size _____ FFD _____

	Exposure 1	Exposure 2	Exposure 3
mA	_____	_____	_____
Time	_____	_____	_____
mAs	_____	_____	_____

Dates																	
Darker 6																	
5																	
4																	
3																	
2																	
1																	
Control 0																	
1																	
2																	
3																	
Lighter 4																	
5																	
6																	

- The numbers represent step differences between test films and control film.
- Plot step differences under dates.

Darkroom inspection checklist

Location _____ Date _____
Inspected by _____

Item	OK	Not OK	Comments
Darkroom: • Temperature (10–20 °C) • Humidity (40% to 60%) • White light leaks. • Ventilation. • Clean & tidy. Safelight: • Working. • Filter faded or cracked. • Correct distance from bench. • Leaking white light. • Correct wattage light bulb. • Safe for film handling time. Film hopper: • Firmly anchored. • Earthed. • Damage free. • Works effectively. Bench top: • Earthed. • Damage free. • Clean. Processor: • Working OK. • QA checks carried out. • Clean. Accessory equipment: • Timer. (working) • Chemical mixing paddles. • Thermometer. (working) • Aprons, gloves, goggles. (damage free, clean) • Name printer. (working, clean)			

Specific gravity/temperature graph

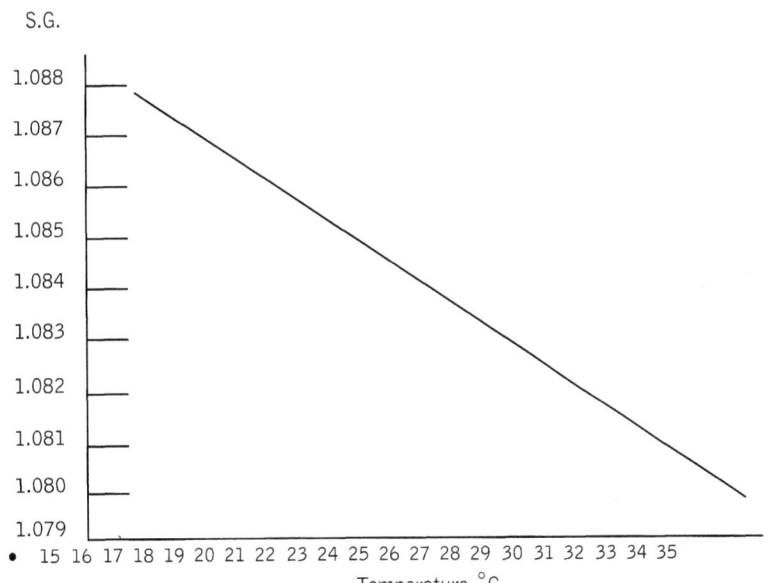

Automatic processor
Daily checks record

Developer: Fixer: Water: Cycle:
Repl. Rate _____ Repl. _____ Exch. rate _____ Time _____
Temp. _____   Temp. _____ Temp> _____   Dryer temp. _____

Two week period:
Commencing _____ Ending _____

DATES											
Dev. temp											
Water temp											
Dryer temp.											
Developer repl. Rate											
Fixer repl. Rate											
Cross over rollers Cleaned											
Circulation Check											
Water flow Rate											
Feed tray Cleaned											
Micro switches Checked											
Filters changed Bi monthly ******											

Remember that filters must be changed **monthly**

_____
_____
_____
_____
_____ _____

Automatic processor
Checklist and periodic replacement of parts

	Replacement parts	Replacement period
a	Springs, gears, shaft retainers, and E rings.	Every year
b	Fuses, temperature controller, safety thermostat, and developer submerged rack rollers.	Every two years
c	Fixer submerged rack rollers, wash water sub-merged Rack rollers, and dryer rack rollers.	Every three years

Inspection period (months)		12	24	36	48	60
Replacement parts	a	*	*	*	*	*
	b		*		*	
	c			*		

(* ... replacement period)

Check list

Rack sections, drive systems
- Rack looseness and geometry
- Rack shaft retainers, gear wear
- Rack helical gear looseness
- Rack roller foreign matter buildup
- Drive motor chain tension
- Abnormal noises

Replenishment system
- Photoelectric detection section function
- Developer and fixer replenisher amounts check
- Replenisher solenoid valve function
- Replenisher pump function

Circulation and temperature control systems
- Processing tank internal foreign matter buildup
- Supply hose solution leaks
- Developer temperature check
- Fixer temperature check
- Circulation pump functions
- Dryer system
- Dryer section temperature check

Miscellaneous
- Finished product photographic quality
- Upper lids, lower lids, and side plate checks
- Water supply and ventilation checks

Processor maintenance checklist

Processor _____ Type _____ Month _____
Developer type _____ Developer temperature _____ Cycle time _____
Developer replenishment rate _____ Fixer replenishment rate _____ Water temperature _____

Date	1	2	3	4	5	6	7	8	9	10	11	12	13	14	15	16	17	18	19	20	21	22	23	24	25	26	27	28	29	30	31
Daily checks																															
Dev. temp.																															
Water temp.																															
Dryer temp.																															
Cross over rollers cleaned																															
Dev. replenishment rate																															
Fixer replenishment rate																															
Recirculation check																															
Water flow																															
Feed tray cleaned																															
Check micro switches																															
Bi monthly checks																															
Dev. filter changed																															
Water filter changed																															
Deep racks cleaned/adjusted																															
Dryer drive belt																															
Drive gears																															
Water flow measured																															

Developer activity chart

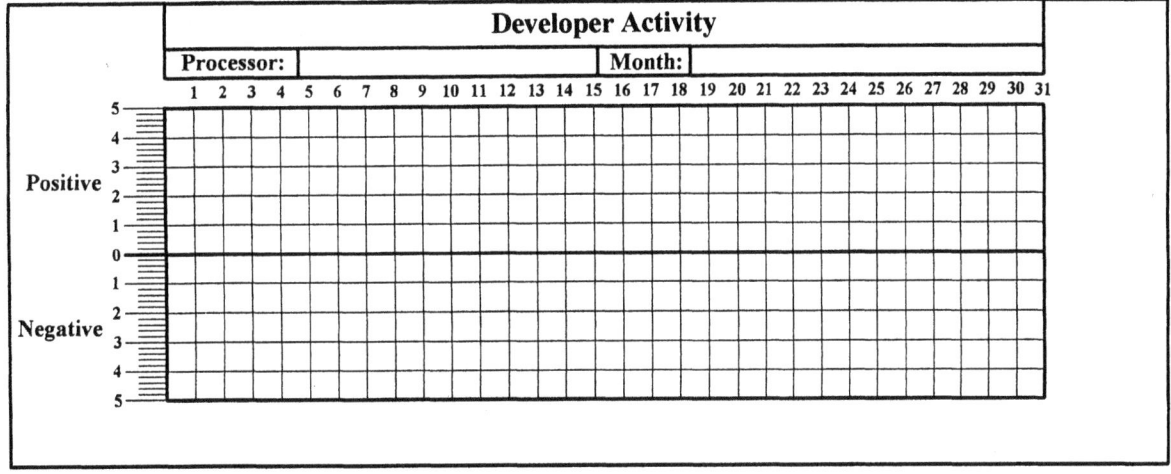

Characteristic curve chart

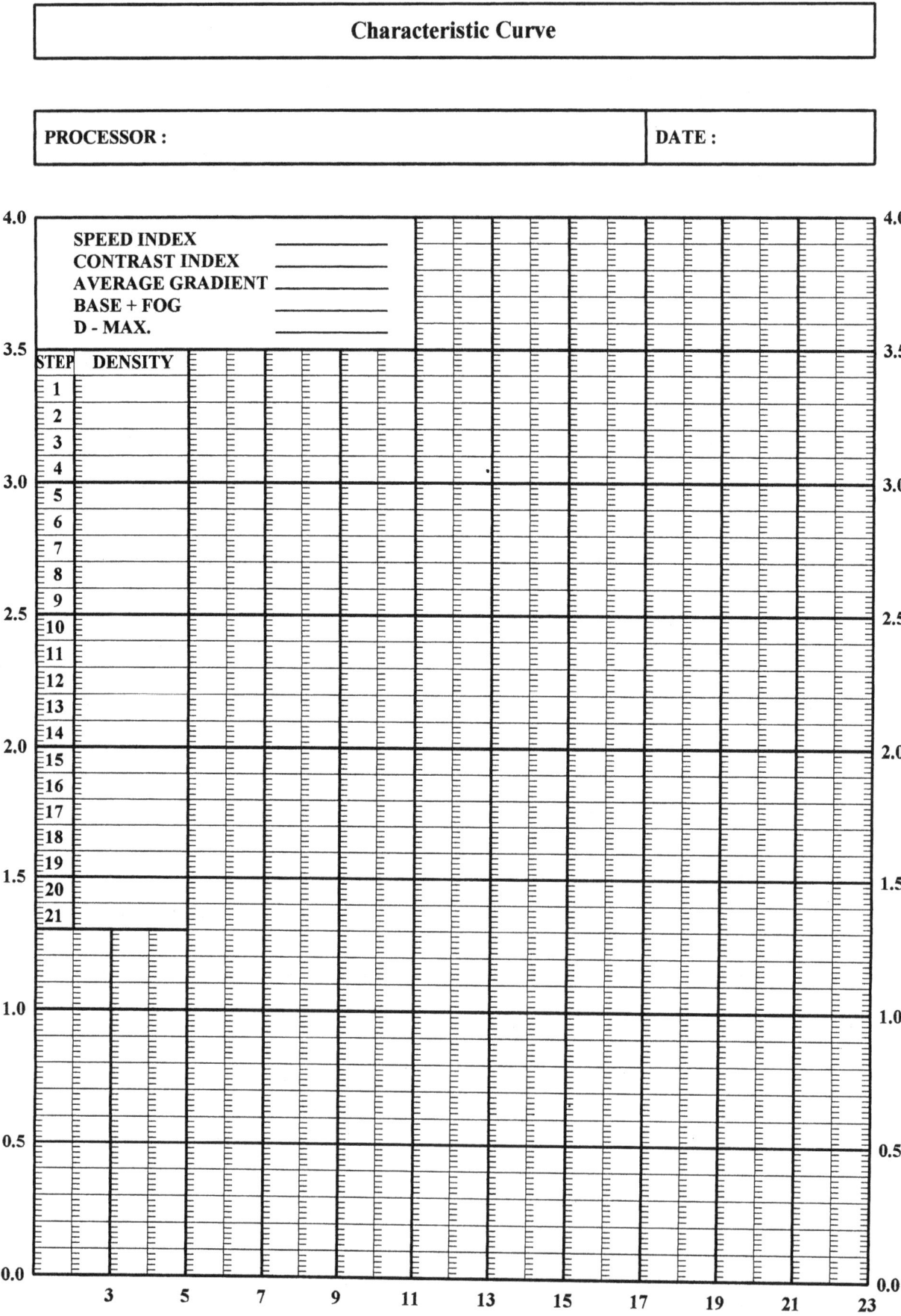

Quality control processing chart

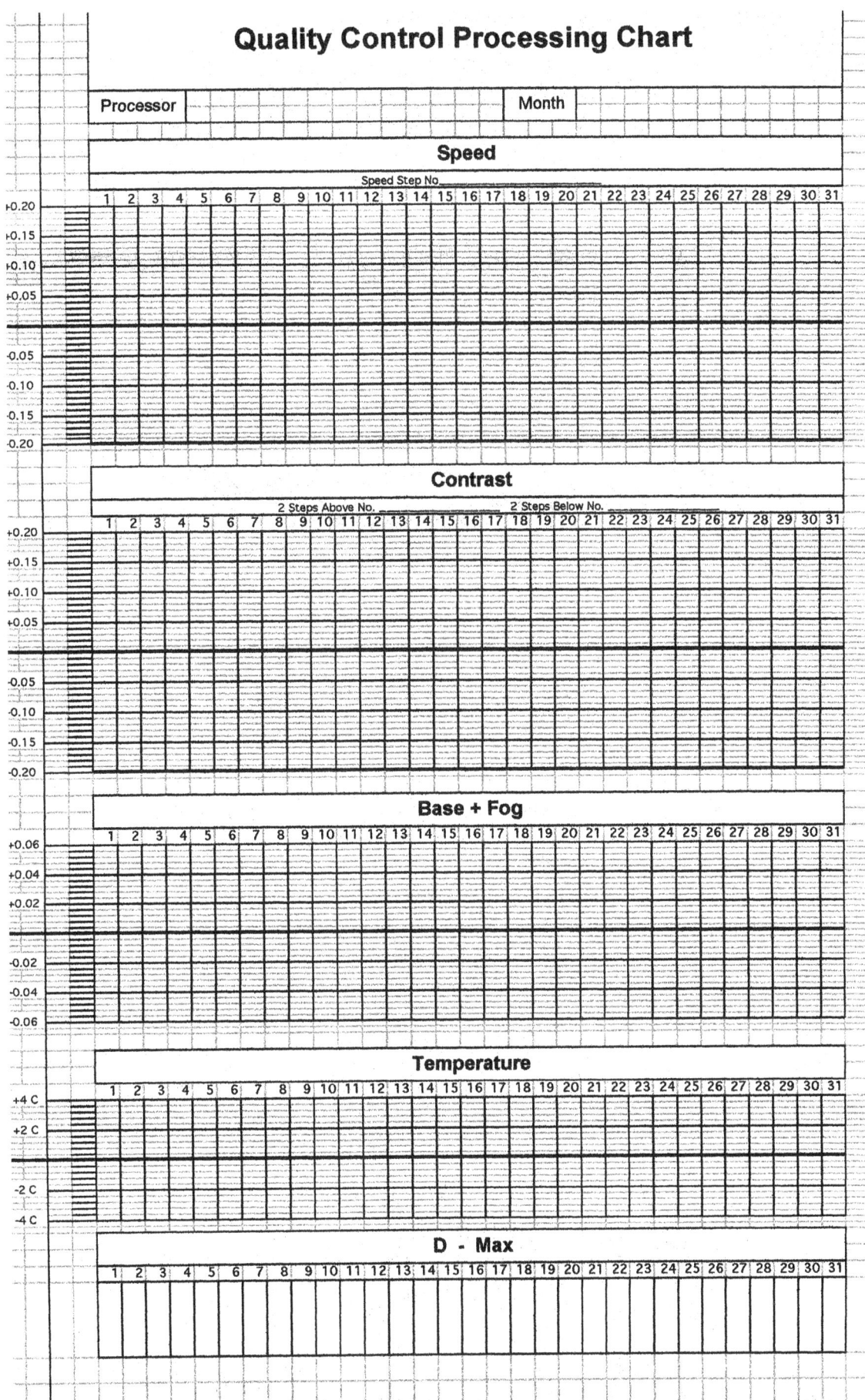

Exposure chart

Examination	Pat. meas.	kV	mA mAs	Sec	FFD (SID) cm	Grid	Screen	Remarks
FOOT								
DP/Obl.					100	—	Detail	
Lat.					100	—	Detail	
CALCANEUM								
Lat.					100	—	Detail	
Axial 40°↑					100	—	Detail	
ANKLE								
AP					100	—	Detail	
Lat.					100	—	Detail	
TIB. & FIB.								
AP					100	—	Fast	
Lat.					100	—	Fast	
KNEE								
AP					100	—	Fast	
Lat.					100	—	Fast	
PATELLA								
PA					100	—	Fast	
Lat.					100	—	Fast	
IS					100	—	Fast	
FEMUR								
AP Lower					100	—	Fast	
Lat. Lower					100	—	Fast	
AP Upper					100	Bucky	Fast	
Lat. Upper					100	Bucky /Stat.	Fast	
HIP								
AP					100	Bucky	Fast	
Lat.					100	Bucky	Fast	
Lat. Horizontal Ray					100	Stat.	Fast	
PELVIS								
AP					100	Bucky	Fast	

APPENDIX B. GRAPHS, CHECK SHEETS & RECORD SHEETS

HAND								
AP/PA/Obl.					100		Detail	
Lat					100		Detail	
WRIST								
PA					100		Detail	
Lat.					100		Detail	
FOREARM								
AP					100		Detail	
Lat.					100		Detail	
ELBOW								
AP					100		Detail	
Lat.					100		Detail	
HUMERUS								
AP					100		Fast	
Lat.					100		Fast	
SHOULDER								
AP					100		Fast	
SI					100		Fast	
Trans Thoracic					100	Bucky	Fast	
Clavicle AP					100	Bucky	Fast	
Clavicle ↑					100	Bucky	Fast	
Scapula AP					100	Bucky	Fast	
Scapula Lat.					100	Bucky	Fast	
CERVICAL SPINE								
AP 1–3					100	Bucky	Fast	
AP 3–7 ↑					100	Bucky	Fast	
Obl.					180	Bucky	Fast	
Lat.					180	Bucky	Fast	
Lat. Supine					180	Stat.	Fast	
CERVICO THERACIC SPINE								
AP					100	Bucky	Fast	
Lat.					100	Bucky	Fast	
Swimmers					100	Stat.	Fast	

THERACIC SPINE								
AP					100	Bucky	Fast	
Lat. Upper					100	Bucky	Fast	
Lat. 10–12					100	Bucky	Fast	
LUMBAR SPINE								
AP					100	Bucky	Fast	
AP L5-S1 ↑					100	Bucky	Fast	
PA					100	Bucky	Fast	
Obl.					100	Bucky	Fast	
Lat.					100	Bucky	Fast	
Lat. L5-S1					100	Bucky	Fast	
SACRAL SPINE								
AP ↑					100	Bucky	Fast	
Lat.					100	Bucky	Fast	
COCCYX								
AP ↑					100	Bucky	Fast	
Lat.					100	Bucky	Fast	
CHEST								
PA					180	Bucky	Fast	
Obl.					180	Bucky	Fast	
Lat.					180	Bucky	Fast	
Lordotic					180	Bucky	Fast	

RIBS								
AP Upper					100	Bucky	Fast	
Obl. Upper					100	Bucky	Fast	
AP Lower					100	Bucky	Fast	
Obl. Lower					100	Bucky	Fast	
STERNUM								
Obl.					100	Bucky	Fast	
Lat.					100	Bucky	Fast	
Lat.					180	Bucky	Fast	
SCJ								
Obl.					100	Bucky	Fast	
Lat.					100	Bucky	Fast	
Lat.					180	Bucky	Fast	

APPENDIX B. GRAPHS, CHECK SHEETS & RECORD SHEETS

PA Close Distance					60	Bucky	Fast	
SKULL								
Townes					100	Bucky	Fast	
AP/PA					100	Bucky	Fast	
Lat.					100	Stat.	Fast	
SMV					100	Bucky	Fast	
FACIAL BONES								
OM					100	Bucky	Fast	
Lat.					100	Bucky	Fast	
Lat. Nose					100	—	Detail	
SINUSES								
OM					100	Bucky	Fast	
PA					100	Bucky	Fast	
Lat.					100	Bucky	Fast	
MANDIBLE								
PA					100	Bucky	Fast	
Lat.					100	Bucky	Fast	
Obl.					100	Bucky	Fast	
T.M.J.'s								
AP ↓					100	Bucky	Fast	
Lat. Obl.					100	Bucky	Fast	

Exposure chart

Examination	Pat. meas.	kV	mA mAs	Sec	FFD (SID) cm	Grid	Screen	Remarks

Post test

Now that you have completed the course your knowledge of the subjects should be much greater. You should now complete this post course test and compare the results with the results of the pre course test.

You will then have an idea of how much knowledge you have gained from the course.

Instructions

This is a multiple choice test. In each question you are given three possible answers.

Read each question carefully.

Indicate the answer that you feel is the most accurate by placing a "X" in front of the letter preceding it.

All questions must be answered

1. Intensifying screens must be cleaned by using:
    a) Plenty of water.
    b) Small circular motions followed by top to bottom sweeps.
    c) Pouring screen cleaner onto the screen then wiping off.

2. Film/Screen contact test tool can be:
    a) Plastic sheet.
    b) Perforated aluminium sheet.
    c) Fine metal mesh.

3. Blue light emitting intensifying screens should:
    a) Be used with blue sensitive film.
    b) Be used only for extremities.
    c) Need an exposure reduction of 10 kV.

4. Grid ratio is:
    a) The ratio of the height of the lead strip to its thickness.
    b) An indication of the focal range.
    c) The height of the lead strips to the distance between them.

5. To test the alignment of the X-ray beam to a grid, the following test tool should be used:
    a) A sheet of perforated zinc.
    b) A lead strip with a series of 1 cm holes in it.
    c) Eight coins.

6. To test a lead rubber apron for cracks:
    a) Use fluoroscopy.
    b) Feel it.
    c) Hold it up to the light.

7. Lead rubber items should be tested every:
    a) Month.
    b) 6 months.
    c) Once a year.

8. Setting up a reject film analysis you should:
    a) Count all the unexposed films.
    b) Count only the film boxes in the film store.
    c) Carry out the analysis without the staff knowing.

9. A reject film analysis could indicate:
    a) Which is the most common fault.
    b) How much money has been wasted.
    c) Both of the above.

QUALITY ASSURANCE WORKBOOK

10. The wattage of a safelight bulb should be, if facing down:
    a) 100.
    b) 60.
    c) 15.

11. A Wisconsin cassette is used for testing:
    a) mA.
    b) Exposure time.
    c) KVp.

12. A spinning top can be used to check the:
    a) Timer of a capacitor discharge X-ray unit.
    b) Timer of a single phase X-ray unit.
    c) Timer of a three phase X-ray unit.

13. Developer temperature should be checked:
    a) Weekly.
    b) Twice a week.
    c) Daily.

14. Sensitometry testing on automatic processors should be carried out:
    a) At 5-0pm every day.
    b) Every Monday.
    c) Before the start of work every day.

15. A densitometer:
    a) Measures the specific gravity of developer.
    b) Is used to make a test film for monitoring film processor performance.
    c) Measures the density of an X-ray image.

16. Empty developer and fixer bottles:
    a) Should have holes put in them and disposed of.
    b) Can be washed out thoroughly and used as water containers.
    c) Can be sold after rinsing.

17. B + Fog stands for:
    a) Film fog caused by scattered radiation.
    b) Film fog caused by light leakage.
    c) Fog present in all film, created in manufacture.

18. A step wedge is:
    a) A foam positioning pad.
    b) A tool for testing tomography cut levels.
    c) A tool for producing a standard set of densities.

19. Consistency of output can be checked by:
    a) Recording how many patients are X-rayed each day.
    b) Making four exposures under identical conditions.
    c) Running a Reject Film Analysis program.

20. A single pulse X-ray unit with a 50 cycles per second mains supply should produce how many dots on a spinning top image with a 0.1 sec exposure time?
    a) 5
    b) 10
    c) 15

21. To test film screen contact use:
    a) Aluminium sheet.
    b) Fast screens.
    c) Fine wire mesh.

22. Specific Gravity level in developer is an indication of:
    a) Its activity.
    b) Its colour.
    c) Its age.

23. Film is most sensitive when:
    a) It is new.
    b) Before it has been exposed to radiation.
    c) After it has been exposed to radiation and before processing.

24. The Step System refers to:
    a) A method of calculating exposures.
    b) How to make a step wedge.
    c) Sensitometry.

25. Clearing time refers to:
    a) The film processing time.
    b) The time the film should be in the wash.
    c) The time it takes for the film to get rid of the "cloudy look" in the fixer.

26. Latitude is defined as:
    a) The range of exposure factors which will produce an acceptable image.
    b) The range of pH values of developer.
    c) The amount of light produced by an intensifying screen.

27. Radiolucent means:
    a) X-rays will pass through.
    b) X-rays will not pass through.
    c) The colour of the light given off by intensifying screens.

28. Quality control refers to:
    a) The number of good films produced over a given period.
    b) Making sure the processor is clean.
    c) A system which assesses and evaluates a particular activity.

29. To make a noticeable change in film density the exposure must be changed by A minimum of how many steps?
    a) 1
    b) 3
    c) 5

30. mAs determines:
    a) Penetration.
    b) Contrast.
    c) Density.

Glossary

Acid A solution with a pH less than 7. It reacts with blue litmus paper and turns it red.

Alkali A solution with a pH greater than 7. It reacts with red litmus paper and turns it blue.

Artefact Marks on a radiograph that are foreign to the image, such as scratches, finger prints or static.

Base + fog The density normally found in unexposed film, caused by manufacture and storage.

Bench top processor A small, automatic film processor, best sited on a bench top. Suited to low throughput.

Beam The beam of radiation produced by the X-ray tube.

Bucky A commonly used abbreviation of the Potter-Bucky moving grid system.

Callipers A device used to measure the thickness of body parts.

Central ray The centre of the X-ray beam. Often used to define the direction of the beam, or, its position, related to a body part.

Cassette A light tight holder that contains a pair of intensifying screens, between which, is placed the film.

Characteristic curve Also known as H & D Curve or sensitometric curve. It is a plotted graph of the various densities of a step wedge image. Any variation in type of film/screen, exposure or processing will vary the shape, or Position, of the curve.

Clearing time The time it takes for a film to lose the cloudy appearance when placed in the fixer during film processing. In other words the time it takes for the unwanted film emulsion to be dissolved off by the fixer.

Collimator A device used to control the coverage of the X-ray beam. Also known as a Light Beam Diaphragm (LBD).

Contrast The difference between the light and dark areas of a radiograph. High contrast is when there are few shades of grey between the lightest and darkest areas of the image. Low contrast is when there are many more shades of grey in the image.

Compression band A strip of material, usually linen or plastic, approximately 20 cm wide, attached at one end to a ratchet device, and at the other to a hook. It is used for compressing or immobilising patients.

Densitometer A device for measuring the density of any specific spot on a radiograph, by measuring the light that is allowed to pass through it.

Density: radiographic Radiographic density is the degree of blackening of a radiograph caused by the deposit of metallic silver.

Density: tissue Tissue density is the mass of body tissue in a given volume, or the concentration of atoms. The greater the tissue density the more X-ray absorption takes place and the lighter the image on the radiograph. (Do not confuse radiographic density with tissue density.)

Detail: radiographic image The amount and quality of information contained in a radiographic image. The amount of detail seen in a radiograph. Is determined by image sharpness, contrast and density.

Detail: intensifying screens The name applied to a type of intensifying screen that gives better image detail, but is less responsive to radiation and therefore requires a higher exposure. Commonly used for extremities.

Development The chemical process of converting the latent film image into a visible one.

Distortion Misrepresentation of a body part outline, in the image, due to changes in X-ray beam/body part alignment or unacceptable object film distance.

Emulsion The active layer of chemical crystals suspended in a gelatine layer, of film, which is sensitive to light and radiation. The word emulsion can also be used to describe the radiation sensitive layer of intensifying screens.

GLOSSARY

Exposure The amount of radiation produced from the X-ray tube, by a pre determined set of exposure factors, kV, mA, seconds. In practice the term tends to be used loosely. The term "exposure" being used to mean exposure factors".

FFD Focus/film distance. More accurately, SID (source image distance).

Focal range The range of focal film distances at which, a grid is designed to be used.

Focal spot The area on the X-ray tube anode where the X-rays are produced.

Focussed grid Grid lines are inclined toward the centre of the grid, to better accommodate the spreading X-ray beam.

Fog Radiation fogging of a film is commonly caused by scattered radiation reaching the unprocessed film. Light fog is caused by unwanted white light reaching the unprocessed film. Base fog is inherent fog, caused in film manufacture.

Filter: safelight A specialised, coloured glass, window, fitted to a safelight, that enables the safe handling of X-ray film.

Filter: X-ray A sheet of metal (usually aluminium) fitted to the port of an X-ray tube to filter out the low wavelength X-ray photons.

Grid A device consisting of alternate radiopaque and radiolucent strips. Designed to allow the primary X-rays to pass through, but absorb scattered radiation.

Grid ratio The ratio of the height of the radiopaque (lead) strips to the distance between them (radiolucent strips).

Grid line The number of lead strips to the cm/inch.

Grid cut off A reduction in grid efficiency due to misalignment of the X-ray beam to the grid.

Hanger Film hanger is a stainless steel frame with clips at each corner, for holding the film when manual processing takes place.

Hazardous chemicals Any chemical which may have an injurious affect. Developer and fixer both fall into this category.

Hydrometer A sealed, weighted glass tube, with a visible scale marked on it, which will float in a liquid. Used for assessing the specific gravity (S.G.) of developer and fixer, the level of which is an indication of the concentration.

Intensifying screens Radiation sensitive screens, placed inside a cassette, on either side of the film. When struck by radiation the screens give off a blue or green light that has a blackening effect on the radiographic image produced. The colour of light emitted depends on the type of fluorescent materials used. Remember that the film colour sensitivity must match the colour of light given off by the screens.

kV Kilovoltage (1000 volts). Controls the quality (penetrating power) of the X-ray beam. Affects contrast of resultant image (high kV—low contrast, low kV—high contrast). Affects intensity of radiation and therefore patient dose, to a lesser extent (mAs has a greater effect on intensity and patient dose).

Latitude: exposure Exposure latitude is the range of exposure factors that will produce an acceptable image.

Latitude: film Is a film emulsion characteristic that increases or reduces exposure latitude.

mA (milliampere, 1/1000th of an ampere) A radiographic exposure factor that controls the intensity of radiation, influences image density and patient dose.

mAs (milliampere-seconds) A radiographic exposure factor. mA x seconds.

Non-focused A grid that does not have focused grid lines.

Oxidation A weakening of developer strength caused by prolonged exposure to air.

Penetration The ability of the X-ray beam to penetrate structures. Determined by the energy of the beam (controlled by kV).

pH Indicates the degree of acidity/alkalinity of a solution. Water is neutral and has a pH of 7. Solutions with pH of less than 7 are acid. Solutions with pH more than 7 are alkaline. Developer is strongly alkaline with a pH of about 14. Fixer is strongly acidic with a pH of about 3.

Phantom An object that substitutes as a patient, when performing X-ray tests. The material used can be anything that is similar in density and thickness to the appropriate body part.

Potter-bucky A moving grid system designed to reduce the amount of scattered radiation reaching the film. Often abbreviated to "bucky".

Primary radiation Radiation emitted from the X-ray tube that has not reached the patient, or object, being X-rayed.

Processing The chemical treatment of an exposed X-ray film that results in the production of an X-ray image. It involves the processes of developing, rinsing, fixing, washing and drying.

Polyethylene foam An artificial substitute for foam rubber. Is ideal for use as patient positioning pads. Can be easily cut to required shapes.

Puffer A rubber bulb with a nozzle for blowing dirt particles from cassettes. Often has a soft bristle brush on the end of the nozzle for the same purpose.

Quality assurance (QA) An overall management program to monitor the various quality control measures being used.

Quality control (QC) The quality control measures used to evaluate performance and outcome.

Quality: radiograph A radiograph is said to have good quality, when the contrast and definition are sufficient to demonstrate all required body parts well.

Quality: X-ray beam X-ray beam quality is determined by its ability to penetrate objects (the higher the penetrating power, the higher the quality).

Radiolucent (radiotransparent) The property of a structure to wholly or partially allow the passage of X-rays.

Radiopaque The property of a structure to wholly or partially stop the passage of X-rays.

Replenisher The chemical solutions that are added to developer or fixer to maintain their volume and working strength.

Replenishment Adding replenishment solutions to developer or fixer to maintain their volume and working strength.

Rinsing In *manual processing* it is necessary to wash off as much of the alkaline developer as possible, before placing the film in the acid fixer. Transfer of excessive amounts of developer into the fixer will result in the fixer becoming prematurely exhausted, lessening its hardening properties and artefacts may appear on the films. In *automatic processing* a rinse bath is not used as the rollers effectively remove the developer from the film and the processing time is shortened.

Scattered radiation Secondary radiation which has been changed in direction from that of the primary beam.

Sensitometry The study of the response of film to exposure and processing conditions.

Sensitometer A standard light source that is used to produce consistent step wedge images for sensitometric monitoring of processors.

SID Source/image distance. Refers to the distance between the source of the X-ray beam (focal spot) and the image receptor (in conventional radiography, the film). Is a more accurate form of, focus/film distance (FFD).

Specific gravity The weight of a substance compared to an equal volume of water. Specific gravity measurement, using a hydrometer, can be used to measure the concentration of developer and fixer.

Stand pipe A pipe which fits into the inside of a manual processing units wash and rinse tanks drain holes. Its height is just below the top edge of the tank, allowing the tank to fill up and drain through the top of the pipe. They are also found in the developer and fixer tanks of most automatic processing units.

Step chart A system of exposure calculation using a series of exposure factor steps. Step charts are available for kV, mA, mAs, time and FFD (SID).

Step wedge Usually made of aluminium, is a block cut to form a standard number and sized steps. Used as a test tool for various quality control tests.

Test tool Specialised items of equipment that can be used to evaluate X-ray or accessory equipment.

Thermostat A device for controlling heat output from a heating unit. Used in X-ray film processing to control the temperature of developer.

Thiosulphate Usually sodium thiosulphate. Is the fixing agent in X-ray fixing solutions.

Timer: darkroom An accurate time clock for timing X-ray film development during manual processing.

Timer: X-ray A device for determining the length of radiographic exposure in an X-ray unit.

Washing All film must be adequately washed to remove the acid fixer and avoid future film deterioration.

Wisconsin cassette A specialised X-ray cassette, used to determine X-ray kVp. Used to check the accuracy of kVp in an X-ray unit.

References

Gray J E et al, *Quality control in diagnostic imaging*, Aspen Publishers Inc, 1982.

Gunn C, *Radiographic imaging, a practical approach*, Churchill Livingstone, 1988.

Hendee W R et al, *Radiologic physics, equipment and quality control*, Year Book Publishers. 1977.

ISRRT, *Quality control handbook*, International Society of Radiographers and Radiological Technicians, 1986.

Jenkins D, *Radiographic photography and imaging processess*, MTP Press Ltd, 1980.

Lloyd P and Warren J, *Basic radiography for general practitioners*, Techsearch Inc, 1982.

McKinney W E J, *Radiographic processing and quality control*, J.B. Lippincott Co, 1988.

McLemore J M, *Quality assurance in diagnostic radiology*, Year Book Medical Publishers, Inc, 1981.

National Council on Radiation Protection and Measurements, *NCRP Report No. 99, Quality assurance for diagnostic imaging*, 1988.

National Radiation Laboratory, New Zealand, *A glossary of physics, radiatioin protection and dosimetry in diagnostic organ imaging*, International Society of Radiographers and Radiological Technicians, 1985.

Quinn H C, *Fuch's principles of radiographic exposure control*, Charles Thomas Publishers, 1985.

Radiation Protection Branch, South Australian Health Commission, *Course notes in basic radiography*, 1996.

Tortorici M, *Concepts in medical radiographic imaging, Circuitry, Exposure & Quality Control*, W.B. Saunders Company, 1992.

World Health Organization, Eastern Mediterranean Regional Office, *Quality systems for medical imaging. Guidelines for implementation and monitoring*, 1999.

World Health Organization, *Basic radiological system, manual of darkroom technique*, 1985.